SOCIAL MEDIA FOR SCHOOLS

Social Media for Schools is a thoughtful guide on how schools can effectively and responsibly use social media to tell their story. Andrea Gribble presents a strategic approach to creating a social media plan that is both effective and achievable for schools of any size. Overall, this book is an essential read for anyone looking to integrate social media into their daily practice.

—**JERRY ALAMENDAREZ**, Superintendent of Santa Ana Unified School District, California

This book is for any educator or administrator looking to navigate the often-overwhelming world of social media. Andrea provides a wealth of practical strategies, tips, ideas, and case studies for using social media to promote student and staff achievements while also addressing the pitfalls and challenges of using social media in an educational setting. Andrea's insights and advice are well-researched and easy to understand, making this book an invaluable resource for anyone looking to effectively use social media to enhance their school's community and communication. Overall, I highly recommend this book to anyone looking to effectively leverage social media to support their school's mission and goals.

—**BRENDAN SCHNEIDER**, Founder of SchneiderB Media and the MarCom Society

What a collection of wisdom and practicality! Andrea Gribble taps her extensive network of school leaders and #schoolpr pros to deliver a master class in school communications and inspire systems thinking. From social media to employee recognition, it's all about relationships. And Andrea walks the talk.

—**JANET SWIECICHOWSKI**, APR, EdD, Fellow PRSA

Andrea has accumulated nearly ten years of experience with social media in schools, and this book is the how-to textbook that every school communicator needs! It's overflowing with resources and real-life examples. Andrea enables novice and experienced school communicators to navigate the world of social media. Read it once, then keep it on your shelf as the "everything I need to know" reference book.

—**JENNIFER BODE**, Mission Advancement Assistant, Arizona Lutheran Academy

When I started my job, I was given a laptop, the password to the school's Twitter account, and an office. I immediately started researching how to accomplish my job of recruiting students, building a strong presence in the community, and managing social media. What I found was #SocialSchool4EDU and Andrea Gribble. I learned how to build systems to manage my time, drive engagement on social media, and track my work to show my value to administration. Without her support, I would be lost for days in the never-ending changes on Facebook, Instagram, and other platforms. This book is a one-stop shop of her best tips, tricks, and strategies for making social media work for your school. Buy it, read it, dog-ear pages, and breathe a sigh of relief. Help is here!

—**CINDY DUNIGAN**, Outreach Coordinator,
Stafford Technical Center, Vermont

Like everything else that Andrea does, *Social Media for Schools* is engaging, encouraging, and essential for any school communicator. It doesn't matter if you are new to the profession or still remember how to work the fax machine, you will find something in this book that will help you celebrate your students and staff, connect with your community, and have a positive impact on school climate.

—**JOHN CASPER**, Communications Coordinator,
Winona Area Public Schools, Minnesota

Social media can be a powerful tool for engaging key stakeholders, if done well. Andrea Gribble has been in the trenches since 2014, working with schools across the country, and she has developed the perfect roadmap to guide school districts in their journey to get storytelling right. She is authentic, relatable, and gracious in sharing both her experiences and her expertise. This book is a collection of great resources, proven strategies, and positive ideas supported by dozens of social media managers working in public, private, charter, and religious schools. So, grab it! This is a must-read for school communicators trying to improve local engagement while gaining a deeper understanding of today's changing social media landscape.

—**JULIE THANNUM**, APR, Senior Strategist for CESO
Communications and a past president of the National
School Public Relations Association (NSPRA)

Andrea's book is great for #schoolPR pros at all levels! From the professional getting started with social media to the veteran, there are great tips and reminders for everyone. It grounded me in best practices and reminded me about the why behind what I do for my district. I appreciated her advice and also loved seeing her personality shine through the text. This is mandatory reading for all social media users and managers.

—**CHRISTY MCGEE**, APR, Director of Communications, Fountain-Fort Carson School District 8, Colorado

Even the seasoned social media manager will find new ideas, suggestions, and inspiration to review, tweak, and implement in their own social media practices. This is a must-have resource! Social media guru Andrea Gribble has been working as a social media storyteller/specialist since 2014. Gribble knows the importance and impact of a good story – especially those that are found within the classrooms and hallways of our K–12 schools. . . . The book provides practical strategies for a successful social media presence . . . all written with one goal in mind: to help school communicators develop and maintain a positive social media presence. Telling your school's story on social media is so important. If you are looking for a place to start, this book is it!

—**DAWN BRAUNER**, Communication Specialist, Portage Community School District, Wisconsin

Andrea is a top expert when it comes to social media in schools. Her book will give you skills to help your schools and school district be at the top of their game regarding social media and marketing. I highly recommend this book to anyone who wants to engage their community internally and externally.

—**JASON WHEELER**, Director of Communications for Garland ISD, TX and national speaker on school marketing and social media

If you're not lucky enough to learn from Andrea Gribble in person, this book is the next best thing. The style and readability make you feel like you're having a personal conversation with Andrea, one riddled with her vulnerable sense of humor. She offers practical next steps no matter where you are in your social media journey as well as myriad examples from the many school districts that have thrived under her tutelage.

— **JILL JOHNSON**, EdD, President of Class Intercom

Social media marketing is a moving target. The strategies, tools, and tips Andrea shares in this book will help you stay focused on that target. Beginner or expert, we all have room to learn from others. This book comes complete with case studies that provide both context and contacts for you to grow in your craft as a school PR professional.

—**JAMIE BRACE**, Director of Communications, Claremore Public Schools, Oklahoma

Andrea's book is really amazing for anyone managing social media at their school. She distills just about everything you need to know to be impactful on social media into easily digestible chapters with highly organized links to so many additional resources. I have been managing social media at our private school for nearly five years and came away with many ideas to effectively refine what I'm doing and ways to do my job more efficiently.

—**JANEEN SORENSON**, Digital Marketing Manager, The Bear Creek School, Washington

There are so many gold nuggets in this book. Andrea's vulnerability in sharing lessons she learned the hard way while highlighting best practices from school communicators around the country make this a must-read for anyone managing social media accounts for schools.

—**JAKE STURGIS**, APR, CEO/Founder, Captivate Media + Consulting

Through her many years of work with school districts and their leaders, Andrea has mastered the ins and outs of social media management for schools. She combines knowledge of the ever-changing space with a helpful and practical approach that works. This is a necessary resource for K–12 leaders and their teams.

—**KRISTIN MAGETTE**, APR, Kansas Association of School Boards, author of *The Social Media Imperative: School Leadership and Strategies for Success*

Whether it's your first or your fifteenth year in communications, you will benefit from reading Andrea's book! The tips and strategies shared are invaluable to any school communicator.

—**BRIAN BRIDGES**, Director of Communications and Public Relations, Lake Hamilton School District, Pearcy, Arkansas

From setting social media policy to determining which platforms to use and why, Andrea shares the secrets to social media success for schools. In my work with small school teams, we see how overwhelming social media can be for school marketers. They have full plates and they're not sure where to start and what to focus on. Andrea's advice will help school leaders build effective systems that save them time and energy (and most likely their sanity!). This is a must-read for small school teams and marketing professionals!

—**AUBREY BURSCH**, CEO, Easy School Marketing

Andrea Gribble is the expert on social media for K–12 schools, and that shines through on every page of her book. She provides practical tips that are easy to execute, coming from a place of empathy and understanding that's critically important in the education space. School communicators are tasked with a lot, often as teams of one, but Andrea comes to the rescue as usual with this fantastic guide.

—**ANGELA BROWN**, Senior Enrollment Insights Leader, K–12, Niche

This is absolutely essential reading for every school public relations practitioner! It's not just a helpful guide – consider it your manual for not just surviving as a social media manager but thriving. Everything you could possibly ever need to make your social media accounts shine is written on the pages of this fabulous book, thanks to the social media cheerleader, guru, expert, and all-around QUEEN, Andrea Gribble!

—**EMILY WHITE**, Director of Communications and Marketing, Sunnyvale ISD, Texas

SOCIAL MEDIA FOR SCHOOLS

PROVEN STORYTELLING STRATEGIES & IDEAS
TO CELEBRATE YOUR STUDENTS & STAFF –
WHILE KEEPING YOUR SANITY

ANDREA GRIBBLE

Social Media for Schools

ISBN: 979-8-218-13820-2

Published by Marketing on the Move
Marketing on the Move LLC
Woodbury, Minnesota

Editorial note from the author on commas, capitalization, and publishing style: I don't want any of you to cringe or hyperventilate while you read this book. Choosing which editorial style guide to follow might have been the hardest decision I had to make. We're communicators, and *The Associated Press Stylebook* is the gold standard for newswriting and PRSA members. But book publishers live and breathe by *The Chicago Manual of Style*. With input from my team of editors and peers, I landed on a hybrid style that perfectly fits the crossover from the world of communication and storytelling to book publishing. And now we can all enjoy the ride – I mean, enjoy the read.

Editing by Emily Rae Schutte and Michelle Rayburn.

Book cover, typesetting, and e-book creation by Michelle Rayburn, missionandmedia.com

I dedicate this book to my loving husband, Bill.

Remember that conversation on our couch? I was convinced I needed to take that job. I couldn't keep chasing this dream. It made the most sense to get the job, and the insurance, and the security.

But you believed in me. You said I could always go back and get a job, but I couldn't just restart this business. You encouraged me to give it all I had for six more months. You knew God had a plan.

I'm so thankful for your encouragement. I love you, and I appreciate everything you have done for me and our family! The best is yet to come . . .

CONTENTS

Foreword by Joe Sanfelippo xv

Why We're Here xvii
How this book will support you and help you create a framework to celebrate your school on social media.

SYSTEMS

Chapter 1: It Starts at the Top 3
Getting leadership buy-in for your goals, how social media benefits your school and community, and gaining confidence to replace hesitation.

Chapter 2: Social Media Policy 15
Establishing district policies, rules of engagement, content guidelines, and policies for staff-run pages. Protecting privacy of students and staff.

Chapter 3: Purpose and Goals 29
Know your end goal for having an improved social media presence, align your goals and practices, and decide how you'll measure your goals. Tips for setting goals for posting about ballot measures.

Chapter 4: Pages You'll Run 41
Get the scoop on district-wide pages, staff-run pages, and social media platforms. Make data-based decisions about your social presence. Learn how to train and support your school-level social media managers.

BRANDING

STORYTELLING

BEST PRACTICES

PROFESSIONAL DEVELOPMENT

FOREWORD

Dr. Joe Sanfelippo

Andrea Gribble walked into my office in 2014. She wasn't lost, but she wondered what was next for her. We talked about schools and stories and branding and all the things we were trying to do to celebrate the great things happening in our schools. In seconds, I could see her passion for her school, but I also saw her passion for celebrating the work that happens in all schools. When we talked about the great things happening in Fall Creek, Wisconsin, she said those things were happening in her town, too, and we both knew that those things were happening all across the country. Yet the narrative of what we do was being told by people who went to school years ago.

We both found it astounding that the people inside the school were allowing the stories outside of the school to become the value of that school. Those stories were actually from so long ago that no one currently in the school was at the school when they took place. When people don't know what you do, they make up what you do, and that becomes the perception of who you are.

Amazing things happen in schools on a daily basis. The moments that take your breath away. The moments that make you want to come back. The moments that help you remember why you got into the profession in the first place. The moments that make you feel safe when you know you let your most prized possession leave your house for the day. Yet, the narrative of what happens in those schools is often based on a negative perception from years ago or a national media perception that schools are failing.

Schools are not failing. Schools take kids from all walks of life, meet them where they are, and help them move forward. Teachers get a group of twenty-five kids, all at different levels with different backgrounds and abilities, and move them – move them academically, move them emotionally, and make them feel safe.

Bus drivers are the first point of contact and set the tone for the day. The kitchen staff gives kids what could be their only real meal of the day. Paraprofessionals give one-on-one attention and brave the ridiculously cold recess times in Wisconsin. Secretaries provide comfort to parents who call and are frustrated or concerned about their kids. Custodians not only keep the building clean, but they see things that not everyone sees because they often walk anonymously with the kids we serve. These are just a few of the job titles and stories that need to be told because when we create value for all who walk the building, we create value for the entire school.

Nobody will change the way *they* talk about school until we change the way *we* talk about school. Andrea and I knew we could help do that for our schools. Having said that, Andrea wanted something bigger. She saw a need, and what she has done to fill the void has been nothing short of incredible. Since 2014, she has worked with thousands of schools, impacting millions of students. Narratives are changing across the country, and she is a big reason why. This book shows how she did it. The work her team did to create a model continues to evolve, but what you will find in the following pages is the toolbox to start it all.

I found myself taking notes, asking our team questions, and making sure we had the process going in the right direction. I also found myself smiling – a lot. The stories are real, her passion is real, and the impact is real. The tools will make you better, and the stories will make you smile. But at the end of the day, the process will do what is most important: helping us change the narrative of schools – one story at a time.

DR. JOE SANFELIPPO
PhD in leadership, learning, and service
Author of *Lead from Where You Are*
2019 National Superintendent of the Year

WHY WE'RE HERE

When you were a child, what did you want to be when you grew up?

A teacher.

A professional football player.

A TV news anchor.

All common answers.

When I was two years old, I wanted to be Dolly Parton when I grew up. That didn't work out – in so many ways!

Did any of you say a school communicator??

I bet not!

But here we are.

And I'm guessing this is exactly why you are reading this book. You've been recently challenged to take on responsibilities that include running social media for your school or district. Or you've been doing the best you can for the past several years but are really struggling with finding great content to share and the time to do it well.

After all, managing social media can feel like a 24/7 role!

You Aren't Alone

When I started doing social media back in 2014, I had no idea what I was doing. Not only did I have to figure out the best ways to set up the social media platforms for the school district and how to schedule posts, but I also had to figure out how I was going to get the content from the school so I could have content to post.

You are *one* person. You can't be in every classroom every day. You have other responsibilities besides social media.

While figuring out how to navigate the ever-changing social media platforms is hard enough, the real struggle is getting a content strategy behind what your school should be sharing. And of course, how to get all of that content.

Social media is a big responsibility. Every word you use and every photo you share is up for public opinion. I mean, we literally want people to comment on what we share.

It can be scary. And you're going to make mistakes. I certainly have, and I'll share lots of them with you in this book.

You are one person, but you aren't alone.

You Need Support

Another concern that I bet you're facing is that you have a lot of people in your school or district who want to tell you how to do your job. Others use social media, so they think they are experts! And you are never, ever going to please all those critics.

Since I was in your shoes not so long ago, the goal of this book is to help provide a proven framework to help you succeed in celebrating your school on social media. You are now the head cheerleader, and we need a plan to get everyone involved. It will take time. However, using this book as your guide, you will be successful!

This book will not break down the specific "how-to's" of the different social media platforms. If I did that, I would literally need to update this book on a daily basis. That's how fast things change. If you need support and training when it comes to that side of your job, you'll want to consider my virtual membership group. I talk more about that in the final chapter of the book. But we have a lot to cover before we get there!

How I Got Here

We all have our stories of how we got into school communications, and since we're going to be hanging out for the rest of this book, I should probably share mine.

When I was thirty-five years old, I was blessed with a layoff.

I can say I was blessed now that it's over, but it wasn't a blessing at the time. I was a single mom with two little girls, and I walked into work on

Monday, November 18, 2013, and was met at the door by the VP of operations. After working in management for nearly my entire thirteen and a half years with the company, I knew what that meant.

I was going to be let go.

After a brief meeting and lots of tears on my end, I discovered my role had been eliminated.

I was shocked. I was scared.

But I wasn't one to sit around. I immediately got to work on looking for another job.

My issue was that in my thirteen-plus years with my company, I had two years as a product marketing manager, two years as an operations manager, three years as a purchasing manager, and three years as a quality assurance manager. Companies were always looking for those magical five to seven years of experience, and I didn't have that in one specific area. I had a broad base of knowledge but not quite enough to place me in a higher-level management role.

My other issue was that I lived in a super small town. There were not a ton of options when it came to finding a company within driving distance.

I started talking with a friend of mine who owned a catering business. Jane needed some help with her growing business, and my broad background was a great fit. After working in marketing for the past several years, I had an interest in social media. Jane said that many business owners are so busy trying to run their businesses that they don't have time to figure out social media.

That's where Marketing on the Move, LLC was born. It was my own official business, where I helped businesses produce content to share on social media to help sell whatever they had to offer.

I worked with Jane's catering business, an accountant, a jeweler, and a photographer. I learned a lot – mostly that every business had a very different type of customer that required a very different content strategy for social media. So that also meant a lot of work for me!

After a few months, the Superintendent of New Auburn School District, Brian Henning, reached out with a simple email. Brian and I attended the same church, and he also was my mom's boss. My mom worked as the administrative secretary. She liked to talk, and she shared how I had lost my job and was trying to start this social media business.

An Open Door

Brian asked if I could help the school start Facebook, Twitter, and YouTube accounts. He had no idea where to start or how to manage it. He just knew that in 2014, social media was becoming increasingly popular and that our small little district was competing with surrounding schools for students. Wisconsin is one of many states with open enrollment options. This means that families could select schools other than the district they physically lived in as long as there were openings in the district of their choice.

I jumped at the chance. I was a graduate of New Auburn, both of my girls were attending school there, and I knew nearly the entire staff. It sounded like the perfect fit.

Plus, it wasn't really like marketing. I was just going to share photos, videos, and stories about the everyday happenings in the school.

My trial program launched on April 15, 2014, and #SocialSchool4EDU was born. I'm proud to say that all these years later, I still manage the day-to-day social media efforts for this K–12 school district of 300 students.

Expanded Opportunities and New Challenges

Right after the kickoff of social media for New Auburn, I was invited to a regional superintendent's meeting to share more about how I was helping them launch social media. It was April in Wisconsin, so of course, there was a snowstorm! Instead of twenty-two superintendents, there were only eight present. But guess what? I signed a contract with my second school!

I wish I could say it was just complete smooth sailing from there, but it wasn't. Supporting myself and my girls was a struggle. I started receiving assistance for food and health care. By August of that year, I was up to my third school, but with hardly any money coming in, I was forced to sell my home and move in with my parents.

Whoa – that was a hard one to swallow. I had been independent my whole life. I built a beautiful five-bedroom home on ten acres at the age of twenty-three. How was it that I now had to burden my parents with two kids and me because I couldn't afford to have my own house?

It was the most challenging time of my life. I really thought I was onto something with focusing on social media for schools, but I also was not yet making enough money to pay the bills. The anxiety was overwhelming, and my therapy sessions weren't enough. I was so afraid.

I decided to interview for a "real" job. I just needed a paycheck.

The weekend I moved out of my home, I received a job offer. It paid $82,000 per year, offered twenty-one days of vacation, insurance – everything I needed!

But accepting this job would mean walking away from what I'd started. Bill, my boyfriend (now husband), sat down with me to talk through the options.

Risking It All

God had bigger plans for me, and I knew it. I decided to lean in for six months and give it everything I had.

That decision is why we are here today!

Now, there have been a ton of ups and downs along the way, but I truly believe that God has me in the exact spot I need to be. I now lead a team of thirty stay-at-home moms who help serve schools around the country. We impact more than 140,000 students and their families each day with the stories we share directly on social media.

In addition to the schools where we manage their social media, I also lead programs that train and support school social media managers. Thousands of schools from around the world have been part of my programs.

I'm Here to Help You!

I'm a natural-born cheerleader. I was a cheerleader in high school, and I continue to do that in this role. Social media managers have a tough job and don't get much appreciation – but I see it! I see *you*. And I'm here for you.

I don't just cheer. I teach. In 2014, after just a few months of doing social media for schools, I started a blog. I have a big mouth, and I'm not good at keeping secrets. So, I started to blog about what works when it comes to social media in schools. In 2020, I started a weekly podcast where

I interview school social media managers from across the country. It's called *Mastering Social Media for Schools*. I share all the secrets! And I'll include a lot of them in this book.

My best advice for others to succeed in life is to hang around really smart people! I've made friends with leaders from around the country who are seasoned school communicators and learned from them. They're top leaders in their field, and you're actually going to meet many of them in this book!

You'll discover that I'm huge on practicality. I don't bother with the fluff. Yes, you need a strategy. But it's more about how to "get 'er done!" I've trained thousands through my blogs, webinars, video tips, boot camps, podcasts, and membership group. I started an email newsletter back in 2015 and am committed to sharing tips and tricks to make your job easier.

Throughout this book, I will often mention school districts. Please know that I have worked with public schools, private schools, charter schools, faith-based schools, and virtual schools. Each school has differences that make them unique, but the overall social media strategy remains similar. No matter what kind of school you work in, the content in this book will help your efforts.

Now it's time to sum it all up in one handy-dandy guide. This book is all of my work so far, packaged up to truly help you!

Ready, Set, Go!

Are you ready to get started? I hope you'll join the number of people who have found content from #SocialSchool4EDU helpful in their work as school communicators. Here are a few that I hope will inspire you to turn the page and get to work.

> "Your podcast with goals for the new school year was one of the best podcasts I ever heard. Such inspiration . . . and at the exact moment I needed it the most. Energy, grace, joy, delegate, boundaries . . . every one of these speaks to me, and I have set them as goals."
>
> —**CHARLENE PAPARIZOS**, Communications Director – Education Service Center of Northeast Ohio

"Andrea is responsive, authentic, energetic, connected, resourceful, influential, and the queen of accountability. She is an encourager focused on finding what's working and sharing it with the world. Mix that together, and you've got the perfect coach. This podcast is my Monday morning motivation! It's invaluable PD fueled by accountability and creativity. The ROI is something even the Shark Tank panel would take a chance on!"

—**JAMIE BRACE**, Communications Director,
Claremore Public Schools, Oklahoma

"I have learned a WEALTH of information, not only here on Facebook (shoutout, *epic* 5-day Challenge!), but also on her podcast, the email newsletter, and website. They are generous with tips/information. It's always easy and intuitive to understand."

—**CALA SMOLDT**, Communications Coordinator,
Sherrard School District, Illinois

"I have tried some other school PR groups, but this is the only one I will continue to subscribe to every year. I am a better communicator because of you and your team, and I have a better attitude. The value of what you provide is worth more than any subscription price. Thanks for being so positive and inspiring."

—**CASSIE GREGORY**, Communications/PR Coordinator,
Coldspring-Oakhurst Consolidated ISD, Texas

"I want to thank you for everything you and your team have done to support, educate, and cheerlead school communications people like me. To have consistent support, encouragement, and empathy from you and the community that you have built has made a major difference in my skill set, my confidence, and my mental health. I can't thank you enough. To have someone like you showing us the way, showing us that it is OK to fail at some things, and showing us that you just need to keep moving forward is inspiring."

—**JOHN CASPER**, Communications Coordinator,
Winona Area Public Schools, Minnesota

BONUS RESOURCES

Throughout the book, you'll find mentions of free bonus content links. Be sure to get those resources here!

SYSTEMS

IT STARTS AT THE TOP

1

Ilt takes patience and planning to win over leaders who have reservations. However, you can prove to your leadership team that social media is a necessity for your school.

Case Study

Meet Jerry Almendarez: Jerry is the superintendent of Santa Ana Unified School District (SAUSD) in California, a district of 41,000 students. He has twenty-eight years in education, serving as a math teacher, principal, administrator, and more. Jerry is a lead-by-example school leader who started using his personal social media platforms early on to amplify the great stories from within his schools. He describes his action plan as "leading with social influence."[1]

The challenge: When he came to SAUSD in the middle of a pandemic, the value of social media increased for Jerry as his team worked to communicate with the community. This also brought a new challenge: he was a new superintendent in a district he didn't know well, and he was trying to tell stories in isolation. He needed to have more voices in the mix.

The process: Jerry's former executive assistant, Lynnette White, came to him with a proposal for a Social Media Ambassador program. Instead of one superintendent trying to push out content he thought was good, they recruited eighty volunteers from the different school locations to train with Andrea Gribble and learn how to tell their school stories.

The outcome: The ambassador program picked up momentum and sparked a wave of interest. Classroom instructional assistants, vice principals,

CASE STUDY

principals, and more started showing the amazing talent and activities at the schools through their lenses. The ripple effect has been compounded because each individual has their own network of connections. These ambassadors share so much more than a communications department can cover on its own. The district website contains links in a directory with all of the schools listed.

"The biggest challenge is helping people really understand the value of a social media presence," Jerry said. He saw the buy-in grow over time as people saw the value of the program. This buy-in experience wasn't new for Jerry. He was confident this time because he'd seen it at work before. When he started a personal professional profile while working at a previous school, he said it made some in the communications department nervous.

Jerry said, "Social media provides an opportunity for the school to create a vulnerable environment that opens up the ability for communities to trust what the educators are doing at that school site, or what the district is doing. . . . It really benefits both the community and the district when parents feel comfortable about what's happening in the classrooms or at a school site."

"If we're not telling our story," Jerry said, "someone else is, and it may not be the version that we want out there." Jerry's results back up the benefit of becoming social influencers on your own story.[2]

It's tough when people don't think your school should have social media accounts. They may not think it's important, are scared of what someone might comment, or just feel like it's a waste of time.

Have you ever dealt with someone like this and struggled with the words to use to convince them otherwise?

I've been helping schools since 2014 with social media and have been in front of DOZENS of school board meetings with at least one member who doesn't believe in it. I also receive questions all the time. Here is an example of one that came to me:

> Hi there! I am a new PR Assistant, and one of my districts is not on board with social media. They won't have a Facebook page because it allows comments, so they have a group instead (because they can turn comments off), which is not working. I obviously feel that the pros far outweigh the cons, and I am curious if you guys have any resources/suggestions that would help me change their minds.

If you've experienced similar hesitation, you understand. I doubt if any one blog or webinar would change the mind of a hesitant leader, but helping leaders understand why it matters, what it means, and how it benefits your community will give you answers to some of the questions and objections.[3]

Why Social Media Matters

I don't watch the news. I hardly watch any television. Do you know how I keep up with the world? Social media!

I don't even follow news channel pages. I just listen to what others are talking about and can then choose to look up more information from there.

I know I'm not the only one. The latest Pew Research data says that

72 percent of US adults use some type of social media, and 70 percent of Facebook users visit the platform every day.[4]

Social media is an efficient and effective way to reach people within your school and in the community. The best way for me to illustrate this is with a testimony from a school that has experienced both the positive and negative effects of social media.

MAKE THE POSITIVE LOUDER THAN THE NEGATIVE

Prior to partnering with #SocialSchool4EDU for social media management, Piqua City Schools in Ohio found their efforts were "all over the place," according to **SUPERINTENDENT DWAYNE THOMPSON** in an interview with Hannah Feller for a guest blog post for #SocialSchool4EDU.[5]

Rather than having one district-wide Facebook page, schools in the district had individual pages. One of those schools experienced a story posted on someone else's profile that falsely and negatively depicted a photo shown out of context. Rather than dwelling on failed efforts to set the record straight, the school focused on telling its own stories to drown out the negativity.

Dwayne said,

> By actively focusing on the positivity in our schools, we have seen that our community engagement and feedback mirror this. Since we have one page for the entire district, everything is in one location, and everybody in our community comes to this one place to figure out and feel what it means to be in the Piqua School District, not just the individual buildings.
>
> Not only that, but we have caught the attention of those outside our community, as well. One of our state representatives called me and offered a ticket for a student to attend the governor's State of the State Address.[6]

For Piqua School District, social media became an opportunity to take hold of the narrative. They got ahead of the rumor mill by flooding their channels with great news. Dwayne Thompson shared some advice for schools who are on the fence about social media, "Don't be afraid. . . . My biggest piece of advice is to feature the voices of many individuals – students, staff, parents – on your social media platforms. Try to get direct quotations and comments so that the experiences of many people join together to create a positive story. The most powerful piece is not the district, but the people in it."[7]

Social media is an efficient and effective way to reach people within your school and in the community.

What Managing Social Media Means

If you can boil down the basics to a few key tips, it can help your leadership team get a picture of an organized plan for social media. For example, a list for a school with fewer than 3,000 students could be:

- ✔ One person in charge of posting to school district accounts. Everything funnels through one email to that person.
- ✔ The entire staff submits stories. Faculty and staff send one to two items per month for social media.
- ✔ Scheduling posts – at least two per weekday on Facebook.
- ✔ Listening, responding, and reacting to comments and questions.
- ✔ Content is monitored every day, and there is a set plan for what to do if content goes against policies.

Grab It!

Dive deep into the experiences of Superior Public Schools in Nebraska. Learn how in his first year as superintendent, Mr. Martin Kobza maximized the use of social media to connect with the community. After developing a system that can be used in any size district, they are celebrating student success, recognizing staff, recruiting employees, and more on a daily basis, reaching more than 7,000 people each week, and they have fewer than 500 students in their K–12 school.

You'll find a link to the **"Using Social Media to Increase Community Engagement"** webinar in the free bonus resources page at socialschool4edu.com/book.

How Social Media Benefits Your School and Community

- ✔ The positive will become so loud that the negative becomes almost impossible to hear.[8]
- ✔ It brings attention to the wonderful students in your district.
- ✔ It brings attention to the great things teachers and staff are doing.
- ✔ Community members have buy-in and understand the needs when a referendum, bond, or levy comes along.
- ✔ The more people see positive attention from the community, the more content staff will provide.
- ✔ It attracts families who want to enroll in your district.

WE NEED TO MAKE THE POSITIVE SO LOUD THAT THE NEGATIVE BECOMES ALMOST IMPOSSIBLE TO HEAR

-GEORGE COUROS

HELP FROM THE TRENCHES

I went inside my social media membership group to crowdsource some responses that might help that PR assistant who asked the question at the beginning of this chapter: "How can I change the mind of district leaders?"

> "It's always a risk, and people already have opportunities to post negatively on their own social media, on the newspaper's social media, and on your community's complaint pages. My former superintendent was very against social media, but after discussing with him the negative comments we get about 'not communicating' on parent surveys and communicating trends of our demographics, he let me test it for our high school. After a year of no issues, as well as the family survey at the end of that year that raved about how the communication had improved . . . he bought in. He has since retired, and our new superintendent is ON BOARD!"

—**JAMIE BRACE**, Claremore Public Schools District Communications / Media Relations, Oklahoma

"You can add this lesson learned by one of my local districts about how important it is to be in the driver's seat of your brand. This particular district did not have an official page; however, there was a page created with the district name (and the word district spelled incorrectly) that was updated just enough with school closings, delays, and other info that people thought it was real. There are currently 1,537 followers. . . . One day, the 'page' wrote back saying she was in fifth grade when she started the page and was going to the high school and didn't want to keep it anymore. The moral of the story is to create your social media presence or someone else will do it for you!"

—**HOLLY MITCHELL-BROOKER,** Community Relations Supervisor at Ulster BOCES, New York

"Groups are the way to go to build community, but you need a public page to push information. Why keep your success stories a secret? Social media is just that – social! Comments engage followers, allow you to respond to questions, and dispel rumors. What's not to love? Carrier pigeons, smoke signals, and weekly papers aren't getting it done . . . you have to play in the social space to be relevant."

—**EMILY BUCKLEY,** former Director of Marketing and Communications at La Salle High School, current Marketing Manager with First Student Inc., Minnesota

"People may be afraid of or just don't want to deal with negative feedback in a public forum. You could share with them how you can block words and a list of responses to negative comments. Be sure to share the policy you will post on your page. Facebook feedback can be a good thing! . . . One more thing that I learned from Andrea and her team: Facebook and other social media allows you to tell your school's story. Every day it is being written within the walls of the buildings. Social media gets it out to the public.

—**REBECCA SCHNEIDER,** Communication Director at Harper Creek Community Schools, Michigan

Create a Strategy

As you seek buy-in for leaders, it helps to have a simple strategic framework.

Content: There are amazing things happening in your school every day. These stories are just waiting to be captured and shared! I'll give you all sorts of ideas for content in this book. If you have a strategic plan or communications plan, you should definitely tailor your content to cover the priorities you have identified.

Consistency: Post consistently at least twice per day on Facebook and at least twice per week on Instagram if you have it. If you can only post once a day when school is in session, start there.

Community: Respond to comments and encourage school leaders to share and comment on posts. Use a school hashtag and reshare content others post with the hashtag.

We'll cover these areas in detail throughout the book, and you'll receive tips on additional strategies and practical ideas for content. But if you need a place to start with getting that buy-in from leaders, present a simple plan for content, consistency, and community. I can promise you that the results from these three keys are proven!

Data and Statistics

I had a call with Steve, a superintendent for a school of about 1,000 students. His school had a Facebook page, but it just wasn't where it needed to be to truly engage the community. They posted one to two times per week.

I wanted to help them. But Steve didn't yet have enough support on his school board to go forward with a proposal from our team. How could we help Steve illustrate the importance of social media for schools to his school board?

Data.

The statistics are always changing, but gathering data is always helpful. For example:

- ✔ The vast majority of Americans, 97 percent, now have a cellphone of some kind. Of those, 85 percent have smartphones.

That's up from 35 percent in 2011.[9] This means that it doesn't matter if they have internet at home or not. They have a cellphone that they can use to access social media along with other internet sites – and it's all in the palm of their hands.

- ✔ Today, 72 percent of the public uses some type of social media.[10] That has increased from 5 percent in 2005 and more than 50 percent in 2011. Facebook is the most popular platform, with Instagram coming in second.
- ✔ Social media is part of a regular routine for many people. Seven in ten Facebook users visit at least once a day. Half of those visit it several times per day.[11]
- ✔ About a third of Americans get news from Facebook,[12] so why not make your school one of those sources?

The numbers have changed since these, I'm sure. So, if you're looking for key data to show leaders, you can go to a source such as Pew Research or Gallup for current numbers to present.

When you consider the habits of the people in your community, social media strategy is not a want but a need for your school.

Get a Seat at the Table

Superintendent Joe Sanfelippo of Fall Creek School District in Wisconsin was an early adopter of social media, and he was my first mentor when I started my business back in 2014. In a podcast interview with me on *Mastering Social Media for Schools*, he explained what social media means for the return on investment (ROI) for a school district.

The ROI is in social capital. Without question, it's in social capital . . . it allows us to be in a spot where when things don't go the way that we want them to, we've developed enough social capital to make sure that we offset whatever ended up happening.

And the more that we do that, the better chance we have to really continue to move things forward. Because if you're not in a position where you're developing social capital with your group, then the minute something goes wrong, they start to question everything about your group.

And one of the things that you [Andrea] taught us about the work that happens here is that people need to make sure that they never forget that *social* in social media matters.[13]

You'll find a link to the full podcast interview from *Mastering Social Media for Schools* in the free bonus resources page at socialschool4edu.com/book.

Dr. Sanfelippo said his district's audience is bigger than the 825 people in a building at the school or the 1300 people in their village. "Our audience turned into the world," he said. "And when our audience turned into the world, it shed a lot of light on the incredible things that our people are doing." Teachers in Fall Creek get calls from all over the country about the work they are doing, which has given them a platform to talk about what they are passionate about.

The ROI in social capital means there wasn't pushback when the district made big investments in projects such as Chromebooks or Wi-Fi on school buses because the community knew why it did what it did.

When Fall Creek started telling the stories of its students and staff over a decade ago, there were sixty students coming into the district by open enrollment and forty-five students going out of the district. That wasn't bad for the budget – netting fifteen students. However, a few years ago, that net number was up to ninety. Eighteen percent of its 825 students are open-enrolled.

Dr. Sanfelippo also said that through social media, the district has been able to feature the culture at Fall Creek, which has helped with its goal of recruiting superstar teachers. "Creating a great culture starts with making sure that everybody's story has value," he said. When it has value, then they are "willing to talk about their story." He said,

When they talk about their story, that builds a culture where people want to be. . . . It's really about building a culture of storytellers, where people are willing to tell the story about the great things that they're doing. Because when they feel comfortable telling that, it means you've created an environment where they can be safe to talk about the great things that are happening. . . . Now you've created a culture of storytellers that people want to be part of.[14]

For leaders who are unsure about using social media, Dr. Sanfelippo said people will talk about the work happening in your school, whether you talk about it or not. He equated social media to having a seat at a table. "Wouldn't you want at least a seat at that table if you knew you were going to be talked about? . . . Nobody's going to hear you from across the room. Sit at the table."

The only way we can influence the conversation is if we sit at the table. That table is social media.

YOU'VE GOT THIS!

- ❑ Help your leaders understand the incredible benefits of having a presence on social media.
- ❑ Be prepared with a summary of how social media can benefit your school and community.
- ❑ Show a simple strategy for how you can get started with regular posts.
- ❑ Use data to show the importance of being on social media.
- ❑ Be aware that people are talking about your school, even if you aren't active on social media officially.
- ❑ Embrace the social ROI that comes from celebrating students, showing the community why your work is valuable, and recruiting superstar teachers to your team.

SOCIAL MEDIA POLICY

2

Thinking of policies may sound stressful, but with a little planning, your school, staff, and community will appreciate how a written policy protects them and promotes positivity. Plus, it offers support and peace of mind for you!

Case Study

Meet Lana Snodgras: Lana is the director of communications and oversees the strategic communications and marketing for the West Plains School District in Missouri. She is an award-winning public relations strategist and writer with extensive professional experience in the public education sector. She has been working in communications and marketing for more than eighteen years. Lana previously spent eight years in radio marketing. She received her degrees from Missouri State University and William Woods University.

The challenge: When Lana first started managing social media for the school in 2010, social media hadn't yet grown to the scale it is now. Around 2013, teachers started asking to have their own social media pages, and coaches were moving toward athletics pages. Lana felt the need to put some guidelines in place for staff and parents. She especially wanted a way to track it all so she could stay aware of what was being posted. Lana also wanted to have a plan in place for how to handle comments and interaction on the school pages.

The process: Lana created policies and a contract for faculty and staff, using ideas from other school policies and personalizing and innovating to make something that fit her needs. She reviews the policies with the faculty and staff at the beginning of every school year. For anyone wishing to start a new page, she has a request form they fill out with the name of the sanctioned page and what they plan

CASE STUDY

to use it for. That is signed with a building administrator's approval before it comes to Lana to have the tech team set up access. If any pages pop up that were created without the proper channels of approval, she reaches out to get a contract signed.

The district also has a social media engagement policy for commenting and behavior expectations for followers. That policy includes what to do if any negative comments or inappropriate posts pop up on pages.

The outcome: For the most part, comments and engagement on social media are positive, but policies give staff and administrators the peace of mind to know there is a plan in place. Parents know what to expect because the West Plains School District Social Media Rules of Engagement content is easy to find on the school website. The school board adopted the policies in 2013, and every superintendent since has been on board with the policies and process. Trust is high between Lana and the faculty and staff because having contracts and policies has provided assurance, support, and boundaries. When they know the expectations, the outcome is positive district-wide.[1]

To hear a podcast interview with Lana Snodgras on *Mastering Social Media for Schools* and to get samples of the West Plains School District policies and forms, check the bonus resources at social-school4edu.com/book.

Have you examined your school's social media policy lately? With social platforms constantly changing and new apps emerging, having a school social media policy is a priority. You may not know where to begin – but that's OK! I'm here to help.

If you already have a policy, when was the last time you reviewed it? Those rapid changes to programs mean your current policy is likely outdated.

Your district policies should include expectations for staff personal use of social media, engagement rules for the community, content guidelines, and processes for staff-run pages. From official representatives of your school to community members who comment, working together will build strong relationships throughout your online communities.

Rules of Engagement

You welcome thoughts and comments on your pages, and you want to encourage engagement. But how can you be intentional about keeping it safe and authentic? When you clearly communicate expectations in a place where your staff and followers can find them, your rules of engagement provide a framework for positive communication.

Social media guidelines provide the information people need to make the right choices on social media. Every person who comments can become an ambassador for your school! Guidelines instruct people on what to do and how to act. You also need policies to outline the repercussions for breaking your rules.

Post your rules of engagement on the school website in a place that is easy for people to find and add the link to your social media accounts. Having this available gives you a foundation for making decisions if you need to remove or hide a comment and a source to point to if you need to explain your actions. Let's look at an example from the New Auburn School

District Social Media Commenting Guideline, which is on the school's Facebook page:

- All of our social media pages will focus on celebrating and supporting our schools, students, and teachers, as well as sharing important news and communicating event information. We encourage you to share your support, connect with other supporters, and visit frequently for news and updates.
- While everyone is welcome and encouraged to comment, our first priority is to protect students, staff, and community members. Comments and/or posts that do not follow this Comment Policy may be removed.
- We have a zero-tolerance policy for cyberbullying and/or posts or comments that are political, racist, sexist, abusive, profane, violent, obscene, spam, contain falsehoods, or are wildly off-topic, or that libel, incite, threaten, or make ad hominem attacks on students, employees, guests, or other individuals. We also do not permit messages selling products or promoting commercial or other ventures. You participate at your own risk, taking personal responsibility for your comments, your username, and any information provided. We reserve the right to delete comments or topics and even ban users if needed. Please be aware that all content and posts are bound by Facebook's Terms of Use.
- New Auburn School District encourages user interaction on its social pages but is not responsible for comments or wall postings made by visitors to the page. Additionally, the appearance of external links, as posted by fans of this page or other Facebook users, does not constitute endorsement on behalf of New Auburn School District. In most, if not all, cases, external links posted by fans will be removed.
- You should not provide private or personal information (phone, email, addresses etc.) regarding yourself or others on this page. Any posts or comments containing personal information of this nature will be deleted.
- If you have questions, please email [school administrator email][2]

Here is an additional example from West Plains School District Social Media Rules of Engagement and posting guidelines:

1. The West Plains School District encourages interaction from Facebook users but is not responsible for comments or wall postings made by visitors to the page. Comments posted also do not in any way reflect the opinions or policies of the district.
2. People making comments on the page are requested to show respect for their fellow users by ensuring the discussion remains civil, especially since Facebook allows individuals 13 and over to join. Comments are also subject to Facebook's Terms of Use and Code of Conduct.
3. Remember that your name and photo will be seen next to your comment, visible to anyone who visits the page.
4. We reserve the right, but assume no obligation, to remove comments that are racist, sexist, abusive, profane, violent, obscene, spam, that advocate illegal activity, contain falsehoods or are wildly off-topic, or that libel, incite, threaten or make ad hominem attacks on students, employees, guests or other individuals. We also do not permit messages selling products or promoting commercial, political or other ventures.
5. Facebook encourages all users to utilize the "Report" links when they find abusive content.
6. West Plains Schools students and staff are governed by the district's Acceptable Use Policy (AUP) when using school owned technology or equipment.
7. If you have questions about the West Plains Schools Facebook Page, please e-mail [address for Director of Communications].

We welcome your comments as a means of sharing your own experiences, suggesting improvements or chiming in on the conversation. To keep our page focused, we have set some comment guidelines.

- This page is moderated and all comments are reviewed by the West Plains School District Communications Director.
- To ensure exchanges that are informative, respectful of diverse

viewpoints and lawful, we will not allow comments that are or include:

» Off Topic. We will delete comments not related to the subject of the page entries.

» Spam. Comments focused on selling a product or service will not be posted.

» Personal Attacks. If you disagree with a post, we'd like to hear from you. We do ask that you refrain from personal attacks or being disrespectful of others.

» Illegal. Laws that govern use of copyrights, trade secrets, etc., will be followed.

» Language. Comments including but not limited to: profane or provocative language, hateful, racially or ethnically offensive or derogatory content, threats, obscene or sexually explicit language will be deleted.

» Links to outside websites. We will not allow fans to include links to websites for any purpose. [3]

....

Now, you might be asking what happens if someone violates something written in your policy. You should be able to take action by reaching out to the user to alert them of their violation. If you want to take further action by hiding, deleting, or blocking any users from your content or pages, you'll want to talk with your legal counsel. I am not a lawyer, and I never played one on TV. It's always best to get them involved if you are unsure of what actions you can take.

Comment and conduct policies cover what happens on your district pages, but what about non-district-related use of social media? You'll need to address this specifically with your staff so they understand expectations, too.

Staff Personal Use of Social Media

It has never been easier to voice your opinions to a large audience. Social media makes it so convenient! That makes social media a very powerful tool. But it also means another potential headache in your job as a school leader.[4]

Kim speaks for many of us with a question she asked:

> Hello, everyone! For our opening day, my superintendent has asked me to talk to our staff about social media. We recently had two staff members post comments on their personal pages that caused parents to call us and complain instantly.
>
> What suggestions do you have for talking to staff about being careful what they post on their pages? I know this will strike up the old debate about their right to post whatever they want on their personal sites. How should I stress the fact that we are held to a higher standard in the education world (even if it stinks) but not come across as punitive or preachy? Every tiny thing is such a big deal to everyone right now, and I just want to protect them from themselves![5]

I totally understand Kim's situation. It's a fine line that we walk, but we need to provide some guidance to our staff. An incredible guide from Dothan City Schools in Alabama gives us a simple starting place. Public Information/Relations Officer Meagan Dorsey even shared the Canva template for others to adapt this for their school district. This is how her guide begins:

> The community expects that educators and staff members be keenly aware that their actions reflect on the profession and the schools they represent. They also expect DCS employees to demonstrate the highest level of professional judgment. We hope you use this helpful guide in making responsible decisions on social media.

The guide then breaks down sections of the document, including:

- Set your level of privacy
- Know copyright laws
- Remember which account you're using
- Don't use the internet to vent
- Remember the Alabama Code of Ethics
- Network with peers
- Fact-check your posts

The guide ends with this simple statement: "Be smart about posting online! In the end, just thinking things through will help you avoid a lot of trouble. Don't be afraid to post with personality; just remember who could see it."[6]

POINTS TO KEEP IN MIND AS YOU TALK WITH STAFF

Other school social media managers weigh in on these takeaways:

✔ Use words such as "we" and "team." **ASHTON** shared a snippet of their guide that says, "It's important to remember that as employees of Danville Schools, we all represent each other. That means that our actions and words as individuals reflect on us all."

✔ Shape it as asking for help. **EMILY** suggested framing it like this:

- It would really help support our students if you . . .
- It would make such a difference if you . . .

✔ Remind your staff that people are watching their pages. **ANTHONY** shared, "There are people on social media who are actively looking for public employees who post controversial things. Those people take screenshots and send them to school boards, superintendents, and communication directors with the goal of getting the employee in trouble. While the efforts rarely result in any major consequences, they can be a distraction. Also, remind them that no matter how many restrictions or fake names they use on social media, their accounts are never really private."[7]

I hope these resources will help you discuss this topic at your school. It can be a touchy subject, but ignoring it will not make it go away. Be proactive with sharing guidelines, and you'll have less of an issue going forward.

TIPS FOR CREATING A POLICY

Here are three things to keep in mind when revising or developing your social media policy.[8]

✔ **Encourage (Don't Scare) Staff** – The goal is to help define the respectful parameters around participating in social media. You don't want to make the policy so difficult to follow that staffers write it off before they even try. Our students are growing up in a digital world. It is very important to teach our young people how social media can be used in a positive, educational way. What better way than using it in school? Here are some ways to encourage staff.

- Emphasize how the policy helps to protect staff.
- Explain how the policy respects relationships between students and staff.
- Give positive reasons for why we post.

✔ **Stress Personal Responsibility** – Whether your staff members are on the clock or not, they represent the school. Here are a few things that are worth restating in your policy.

- Use common sense.
- Be mindful that content you publish will be public for a long time.
- Use your own personal email when setting up profiles.
- Respect the privacy and rights of both colleagues and students.

✔ **Address Employee-Student Relations** – Communication between employees and students is vital to the educational process and experience. With the vast technology available today, it can be advantageous to use social media in the learning process. Having a clear understanding of the expectations around this topic is critical. Things to keep in mind:

- Employees are never under any obligation to accept friend or follower requests from any student. However, if you accept one student, it may be recommended to accept all to avoid favoritism concerns.
- Employees must exercise great care in connecting with students (if it is allowed).
- Employees are responsible for ensuring any relationship and all dialogue with the student is kept professional in nature.

I advise consulting legal counsel before adopting any new policy or procedure related to social media. Samples from other schools are helpful, but they don't replace legal advice.

Grab It!

In the free bonus resources for this book, you'll find links to samples of school policies and a handy template for creating a graphic with your social media for personal use policy. Get those at socialschool4edu.com/book.

- **Sample school policies** from Fairfax County Public School in Virginia and Dothan City Schools in Alabama
- **A free template** to create a social media for personal use graphic provided by Meagan Dorsey at Dothan City Schools.

Could you use support as you tackle the task of creating your policy? The #SocialSchool4EDU membership program includes a community where we provide daily support in peer-to-peer conversations in a private Facebook group. It's a great place to bounce your questions off people who can share ideas and support.

Student Privacy: Opting In or Out

Student privacy is a legal matter that must be followed by every school. I'm going to keep it simple here. I'm just talking about photos and names. Because of the FERPA law (Family Educational Rights and Privacy Act),[9] schools must ask for permission to use student photos and names on the website, in the newspaper, and on social media.[10]

Schools must ask for permission to use student photos and names on the website, in the newspaper, and on social media.

Getting approval to use the photos and names of students happens in two ways – either opt out or opt in.

Most schools operate under an opt-out process. Opt out means that every child is assumed to be allowed to be pictured unless they receive a form from the family saying they can't.

If your school uses an opt-in process, then the process is reversed. You assume no student can be pictured unless you have a form allowing them to be. In that case, you have to collect a form for every student in your school before you can use their photo.

An opt-out process is much easier to manage because you will likely have just a few parents that choose not to have their child's photo used.

No matter which process you follow, my suggestion is to keep your forms simple. I've heard about some schools that have various levels of permission granted. For example:

1. You can use my child's name but not their photo.
2. You can use my child's photo in the newspaper but not on social media.
3. You can use my child's photo in the yearbook but not on any other media.

You can quickly see how this option will make it nearly impossible to manage on a daily basis! Most schools we work with have only a handful of students that can't have their photo shared on social media.

As I said, make your form simple. Either the child's photo can be included in everything or nothing. Some schools even include the yearbook as a part of this permission. This means that if they are opt-out, they opt out of every form of media, including sharing their photo in the school yearbook.

I get great questions all the time from schools about how to handle these sticky issues. It's likely that the same questions have popped up at your school about sharing names and faces. Let's look at a few:

What should we do with photos of students who have opted out?

While it might be tempting to blur faces or use stickers to conceal faces, do not cover students' faces if their family has opted out of being on social media. Choose a different picture. A photo that is obviously edited brings even more attention to those students!

Should we include student names on social posts?

One social media manager named Amber emailed me after a concerned community member reached out to her. You may have received similar questions. It read: "We had an inquiry from a community member (not a parent of a student) pertaining to student names we have on Facebook along with photos of our students. The issue was that they were concerned with human trafficking/kidnapping and potentially giving out information."

The concern addressed that even if students weren't named in the post, they were holding certificates or schoolwork displaying their names. She had a valid question.

Your district should decide early on whether you will include names on social media posts. My philosophy is that if you would list out the names in a newspaper article, you should include them on social media. After all, the newspaper is likely online, so the students' names will already be available for people to see. This would include posts about special recognition and awards. For everyday posts that simply show students in action, we don't include names.

Why not include names on every post? On social media, I hope you're sharing a LOT of photos from your school. On many of our school district pages, we are posting more than three times per day. If we had to include a name for every single photo we posted, we would be in trouble!

When it comes to your district policy on including names, there are many variations:

- ✔ No limits – use names as often as you want
- ✔ Include names (first and last) only on special recognition posts like awards and scholarships
- ✔ Include names (first and last) only on posts about high school students
- ✔ Include only first name and last initial on posts that warrant the students being identified
- ✔ Never include names

What you decide is completely up to your school. It's just important that once you make a decision, you follow it strictly.

You'll need to watch the photos you use to ensure you follow the policy you decide to use. As Amber's email indicates above, some photos show full names on the certificate the student is holding. Other photos include names inadvertently because student names are printed on desks. We've opted to edit those photos by covering up the name so it is not legible in districts where full names aren't allowed.

What if someone else tags a name in the comments?

You can have direct control over what you post as the school, but what if someone comments on the photo and tags someone? First, we never allow tagging directly on the photos we post on our school's social media pages. While tagging might garner more views, we don't recommend it. After all, when someone is tagged directly in a photo, it can show up on their personal Facebook page.

Second, we cannot control all the comments made on our posts. While we need to keep things safe, following the social media commenting guidelines you have set up for your school, we can't control everything people put in the comments. You can hide or even delete those tagged names of students or their parents – but that would be a full-time job in itself!

....

If you don't have a policy yet, I strongly encourage you to tackle this topic. It will make your job much easier as a social media manager for your school!

YOU'VE GOT THIS!

- ❑ Create or review the social media policy for your school.
- ❑ Enlist legal counsel to advise on your policy.
- ❑ Establish rules of engagement for your community and communicate those in a place your followers can access them.
- ❑ Communicate policies and expectations clearly to staff regarding their use of social media.
- ❑ Establish opt-in or opt-out policies for families.
- ❑ Checkup: Are you consistent in applying your policy?

PURPOSE AND GOALS

3

Knowing your district values, sticking to your goals, and having the right tools means you don't have to question what you're sharing on social media!

Case Study

Meet Brendan Schneider: Brendan is the founder and CEO of SchneiderB Media, a digital marketing agency for schools, and the MarCom Society, a community for school marketing and communications professionals. He spent fourteen years working at Sewickley Academy, a private pre-K–12 school outside of Pittsburgh, and held other previous roles in education, one of those being a director of enrollment.

The challenge: Brendan helped Sewickley Academy transition from print, direct mail, and billboard marketing for enrollment efforts to social media with the expectation that it would bring full-pay families knocking at their door. It didn't. He soon discovered that without a strategy, social media by itself wasn't effective. Being present wasn't enough to drive enrollment.

The process: Brendan now encourages schools to think of social media as having a dual purpose: recruitment and retention. Schools do retention well, he said. Retention posts that celebrate students affirm a family's choice of enrollment, but unless followers share it somewhere else, it doesn't have any word-of-mouth effect. They are posts mostly for consumption. In contrast, recruitment posts have the goal of driving someone to one of your web assets – a website or a blog – where you can convert them to a lead.

CASE STUDY

The outcome: Based on the concept in Gary Vaynerchuk's book *Jab, Jab, Jab, Right Hook*, Brendan said schools need to think retain, retain, retain, recruit in how they balance social media posts. Action steps in a recruitment caption might be to sign up for an event, take a tour, or download an inquiry magnet. When choosing content, Brendan advised asking yourself what followers want to see. Being intentional with format and purpose can help increase enrollment interest *and* still celebrate your great school.[1]

NEVER give up
THE OPPORTUNITY
TO SAY SOMETHING
GREAT
ABOUT your SCHOOL!
- JOE SANFELIPPO

Why does your school or district want to be on social media? You have to have a purpose behind your presence on these platforms. If not, why would your school invest the time and energy that it takes to run an effective social media program?

Many schools overcomplicate this step. The purpose of social media is to celebrate your students and connect those stories to your community. You are giving outsiders a front-row seat to the amazing things happening inside your district! We'll talk specifically in chapter 17 about social media metrics, but here we want to cover the impact that reaching more people with your stories can have.

So, what would the end goal be for an improved social media presence? Your goals might have to do with your overall mission for social media (building community engagement, showcasing your brand, celebrating students), or they might be more specifically related to efforts and engagement (metrics, leads, traffic).

The main goals that I encourage you to focus on are related to the following:

- Increase community engagement
- Increase enrollment
- Improve employee morale
- Improve school culture
- Improve communication
- Attract candidates for staff and support positions
- Support the passing of a bond, levy, or referendum

Social media can help with all of these, but it's important to identify which is the most important to your school. Most schools I work with indicate communication and community engagement as priorities. They want to meet their community where they are at.

Did you know that 70 percent of your community has no kids in school right now? This means they aren't receiving your direct emails or phone calls. You may not have a local newspaper, and even if you do, how many subscriptions are there? Social media gives you the opportunity to reach thousands of people every day!

Your goals are also related to your "why." Why use social media?

- To celebrate students
- To shine a light on staff
- To keep the community and alumni informed
- For crisis management
- To own the story
- To lead by example – social media can be POSITIVE!

We'll cover a lot of content in this book that relates to those purposes and how to accomplish those goals.

YOUR SOCIAL MEDIA PLAN

In podcast episode number 12 of *Mastering Social Media for Schools*, I talked about dialing in on a solid plan for social media. The episode includes the following:

- ✔ A writing exercise to reflect on your social media strategy from last school year
- ✔ What social media metrics you should evaluate on Facebook, Instagram, and Twitter
- ✔ A simple way to add more Facebook fans to your page
- ✔ A method to develop your #1 goal for social media

Plus, it includes links to video tips for how to invite engaged fans to like your school's Facebook page, how to reach more people with your posts, and the best times for posting.[2] You'll find a link in the free bonus resources at socialschool4edu.com/book.

Grab it!

Learn the simple secrets behind social media for K–12 schools. I've trained thousands of schools across the country, and now I've packaged my best tips for you to receive in a **three-part video training**! Get the link in the free bonus resources at socialschool4edu.com/book.

Alignment of Goals and Practices

In a guest post for #SocialSchool4EDU, public relations expert Barb Nicol, APR, discussed what we want people to know, feel, or do as a result of our posts. She said, "It's easy to fall into the habit of posting things on social media because you have a good photo, want to promote an event, have established a special posting routine on certain days or need to fill your feed with something engaging. But it's a major opportunity missed if you don't take the time to align those posts with your overall communication goals."[3]

When goals are clear, it will be easier to identify what types of posts will help you reach those goals. Nicol emphasized aligning social media goals with the district's overall values. She said, "If you frame your social media posts so they support your values as your goals, your posts will support your brand and demonstrate who you are as an organization."

If we learned anything from posting during a pandemic it was that being overwhelmed was real. Some schools looked to other districts for inspiration to see what was getting the most attention. However, Nichol advised that what works for one community might not for another. "Make sure your social media posts match your community's reality. Authenticity is key to a successful social media strategy," she said. The other thing we learned during that time was to modify. Nichol suggested that there might be times when our goals related to enrollment, staff retention, voter approval, or community pride seem distant or unrealistic during times of crisis, but pivoting the focus is OK!

> "If you frame your social media posts so they support your values as your goals, your posts will support your brand and demonstrate who you are as an organization." Barb Nicol, APR

ALGORITHM FORMULA

One of the big factors in the reach of your posts is the reactions it gets: likes, comments, shares. But these reactions are not all created equal. If we were to create a scale of impact, here's how it might go:

- ✔ A like is worth ONE point.
- ✔ A comment is worth SEVEN points.
- ✔ If someone shares your post, it's worth FOURTEEN points.

That means hitting the share button is fourteen times more powerful than liking the post. Tell your staff and parents that their interaction matters.[4]

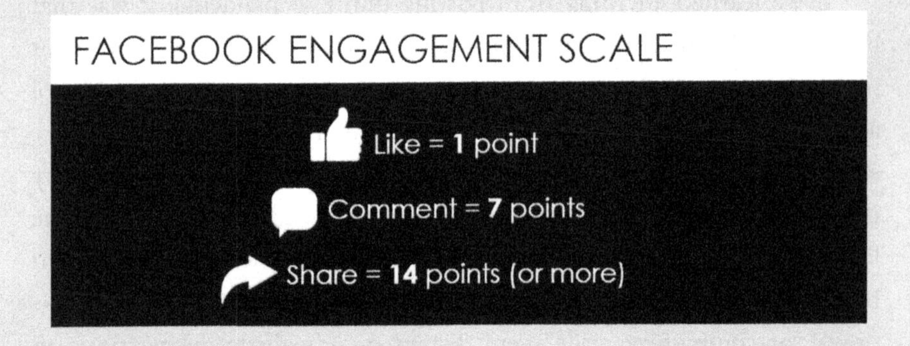

FACEBOOK ENGAGEMENT SCALE
Like = 1 point
Comment = 7 points
Share = 14 points (or more)

PIVOTING DURING THE PANDEMIC

Perhaps you have a referendum in your future or are facing difficult budget cuts. While it may feel tone-deaf to share posts that focus on financial challenges given the recession that is likely to impact the majority of our community members – you can focus on trust, value, transparency and community pride. These all provide a solid foundation upon which to build a financial request or share difficult financial news when the time is right.

Maybe you are trying to recruit certain types of staff or improve staff satisfaction. While it may or may not seem like the time to post positions, you can demonstrate how much you value your staff, how important they are in the work being done and how you listen to what they need.

While it may feel inappropriate right now to roll out an enrollment campaign, you can still showcase what great schools you have and how incredible your teachers are. You can provide parents with extra support, resources, and ideas for how to manage during these crazy times. You can demonstrate that you are the education experts, and parents can trust you to deliver what their students need – in good times and in bad.[5] —**BARB NICOL, APR**

Announcing Ballot Measures

If your district is going to ballot soon or perhaps is in the discussion phase, you might be wondering how often to post, what to post, how to respond to negative chatter, and more. When it comes to social media, you'll have specific goals for a referendum that will need to fit into your overall goals. Our team has helped with more than one-hundred referendum and bond campaigns, and we've learned a few things.

Communication with the public plays a pivotal role in any school district. It is the cornerstone of building relationships with your constituents. The true test of how well you are communicating is often presented when you must ask your community for funds. Whether it is for operational or facility needs, you need to ensure the voters know why you need the money.

When it comes to using social media for your ballot question, you need to assess the climate in your community. We've had districts that have avoided using social media altogether because they did not want to create a forum for the "Vote No" group to shout their opinions. When a school decides to post, they should expect comments and questions from both supporters and opponents. You know your community best and will have to make the decision.

Most of our schools choose to share ballot-related information on social media, but the frequency of those posts has definitely declined. Instead of posting two to three times per week for six weeks leading up to voting day, they may only post about the bond/referendum weekly.

Dorreen Dembski, an experienced school communications professional, shared some of her expert wisdom in a guest post for #SocialSchool4EDU.[6] This isn't the time to start a new communication strategy, she said. "Long before a referendum, a strategic organizational communication plan should establish well-worn communication paths between your school district and your stakeholders. Then, when a referendum comes onto the scene, the pre-referendum information is a special topic plan within an overall strategy."

Dembski said that all communication plans have four key components, represented in the acronym RACE: Research, Analyze, Communicate, Evaluate. As communicators, we're specifically vested in the third point. Here are some examples from Dembski on communication channels we might use:

1. **Fact sheet:** A fact sheet grounds the key messages into one single document.
2. **Website**: Dedicate a section of the website to store all the materials for a voter.
3. **Social media plan:** Ideas for social media posts include "Did you know" posts about your referendum needs and solutions; announcements of where to find information; pictures – lots and lots of pictures pertaining to building plans or how students will be impacted; answers to frequently asked questions.
4. **Videos:** Think about how to incorporate short videos into your overall social media plan.
5. **Other communication tools:** Email newsletters, print flyers, mailers, informational open houses, presentation boards and posters,

frequently asked questions, and an email address where the public can submit questions are all helpful. All communication tools are integrated into an overall calendar and used to organize and implement your communication plan.[7]

You may or may not oversee other communication channels in addition to social media. Dembski emphasized that all referendum information should be consistent across all channels. Collaboration with the leadership team and anyone who has responsibility for the referendum communication is key.

Prepare your team for how to answer questions on social media. Dembski also recommended watching for trends on social media regarding questions about the referendum. This could include listening in parent and community groups on social media. Patterns can provide cues for where additional information needs to be shared. "Remember – your purpose is to inform the public, so simple language and transparency is key," Dembski said.

Pre-Voting To-Do List

We hosted a webinar featuring Heidi Feller, where we covered referendum communication. From that discussion, my team created a list of eleven to-do items to help you:[8]

1. **Have a strong social media presence** – A referendum is NOT the time to start using this powerful communication tool. You should already be celebrating your school on a daily basis. This means posting regularly on your social media channels!
2. **Gather the troops** – Administrators, architects, general contractors, PR people – get these key players in the same room. You all have a key role to play.
3. **Get a communication strategy** – What tools will you use to share information? This includes press releases, mailings, postcards, meetings, website updates, as well as social media posts. The key here is to layer, layer, and REPEAT. Push out talking points on various platforms at the same time.
4. **Branding is key** – Update your profile image and cover image to reflect your upcoming ballot. Create a special FAQ graphic in line with your branding. Recycle graphics used in mailings,

postcards, and on the website. Never pass up the opportunity to use your school colors and logo.

5. **Just the facts, ma'am** – Stick to brief sound bites pulled directly from other communication platforms. Compose social media posts that are easy to digest and understand. If you can't understand the post, then your taxpayers most likely won't, either.
6. **Get creative** – Shoot a short video explaining a specific talking point, consider a Facebook Live Q & A session, record a podcast, or even start a weekly feature such as Twitter Tuesday. Don't forget to always use great visuals!
7. **Business as usual** – Tell your district's story! Referendum posts should be one to two posts per week in the eight weeks leading up to voting day. Every post should not be about the upcoming ballot question. Your daily posts of classroom happenings remind folks of the value your district brings.
8. **Monitor posts** – Be prepared for negative comments and have a plan to respond. This is critical. Social media is a great place to learn what questions may be out there regarding the ballot question. Use the platform as a listening tool, and then dial in your communication strategy from that.
9. **Remind folks to vote** – Your social media followers are likely supporters of the school. Make sure you remind them to vote! You can do this months before through absentee ballot reminders and then definitely on the morning of the vote.
10. **Report results** – Be your own news channel. When you know if it passed or failed, let your community know via social media. Be prepared with statements for both scenarios!
11. **Keep folks updated** – If your vote passes, don't stop communicating! Long building projects can take time, and you will need to update your followers. Now that you have the funds, keep the community in the loop on the use of those monies. If your vote fails, decisions may also need to be made. Communicate those, as well.

Don't wait until your district needs voter input to start consistently communicating with your community and celebrating your students. When the time does roll around for you to request your community's support for needs at the school, they will be there – ready to support with a "yes" on that ballot because they've seen thousands of snapshots of school success.

Grab It!

You'll find a link to a printable PDF of the "**Referendum Checklist for Social Media**" and "**Referendum Communication on Social Media**" in the free bonus resources.

Measuring Your Goals

Now that you've identified your goals, how do you measure them? This can be a tricky part! Again, we can relate it to some of the social media metrics that are available on each platform. That will be discussed in chapter 17.

Earlier, I identified these common goals, and here are some ideas for measurement:

- ✔ **Increase community engagement** – Evaluate reach and engagement on the social platforms. Conduct a survey to have the community rate how well they feel engaged with the school or district. Repeat this survey at least annually to measure feedback.
- ✔ **Increase enrollment** – This is straightforward and can be measured year to year. If you are a public school with open enrollment in your state, I encourage you to evaluate students enrolling in versus students enrolling out to another school.
- ✔ **Improve employee morale** – Conduct a survey to rate employee morale. Repeat this survey at least annually.
- ✔ **Improve school culture** – Conduct a survey with students, staff, and families. Questions can reference inclusivity, friendliness, academic success, and more.
- ✔ **Improve communication** – Conduct a survey with students, staff, and families. Repeat this survey at least annually. Find out how respondents get information, how they'd like to get information, and more.

- ✔ **Attract and retain qualified staff and support positions** – Evaluate applications, unfilled positions, and the time it takes to refill jobs.
- ✔ **Support the passing of a bond, levy, or referendum** – This one is easy, and you will find out after the ballots are counted!

I won't go into detail on survey types and questions here, but there are many options to choose from. You can use online surveys, focus groups, and digital brainstorming. There are companies that specialize in this, but you can also do it on your own. We often share survey examples in our membership group, so check out this valuable option.

When you set goals and implement consistent practices, you'll be ready to pivot whenever change comes your way. And your district values will always be front and center in all you do.

YOU'VE GOT THIS!

- ❑ Review your district values and your communication goals.
- ❑ Evaluate your current social media content to see if it aligns with your purpose and goals.
- ❑ Identify goals for the next term or year.
- ❑ Communicate consistently and celebrate your students.
- ❑ Measure your goals and assess regularly.
- ❑ Create a plan for communicating about your upcoming ballot measure.

PAGES YOU'LL RUN

4

You only have so many hours in your day, so efficiency and effectiveness are top priorities. Here's some reassurance: you don't have to be on all the channels! Let's figure out the best ones for you and then set you up for success.

Case Study

Meet Peg Mannion, APR: Peg Mannion is the community relations coordinator at Glenbard High School District 87 in Illinois in the suburbs of Chicago. The district has around 8000 students in four high schools, making it the third-largest high school district in Illinois.

The challenge: Having four high schools in a district brings up the question of whether each should have its own Facebook page. They happen to be in three different conferences and even sometimes play one another in sports. In a district of 8000 with four schools, can one page work?

The process: The four high schools in Glenbard District 87 share one Facebook page. The district handles Twitter and Instagram differently: the dance team, student council, sports, clubs, teachers who share professional development on their individual accounts, and more each run their own accounts." However, all district-wide posts, including those that celebrate students and staff, go on the Facebook page, as well as the district Instagram and district Twitter accounts. When students from each high school were named National Merit semifinalists, the district shared one post to celebrate all of them.

The outcome: Having one Facebook page provides what Peg calls "one-stop shopping." Information is posted on the page for people in all schools to see at the same time. It saves on confusion and keeps

CASE STUDY

the schools unified. It also allows the district to have a consistent voice in its presence on social media.

There is crossover in communities outside of the school, and having one Facebook page avoids putting people into silos. There is overlap in where they work, in community activities, where families go to church, and in park district summer programs. Peg says this system of posting to one page is ultimately more efficient.

Although Peg is the social media manager for the district and for the Facebook page, posts for the individual clubs and activities on Twitter and Instagram are handled at a school level. Adult sponsors work with students on content approval. While there may be a case for having separate pages for a much larger district, this method has worked well in Glenbard.[1]

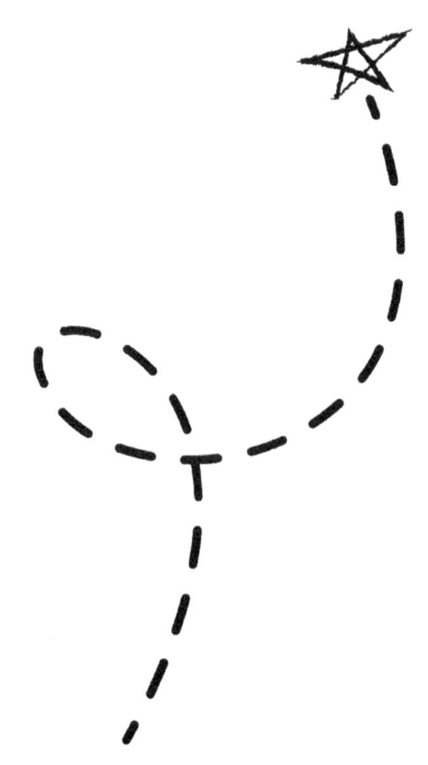

Which platforms should we be on? How often should we post? What should we post about? You already have an overflowing list of tasks and responsibilities, and the answers to these questions will direct your time and budget. Therefore, it's important to choose wisely! This chapter covers a few things to think about before you get started. Already started? No worries! I'll help you think through any changes you may need to make.

Before we talk about anything else, let's talk about districts with multiple schools. I get this question often: Should we have different pages for each school?

District-Wide Pages

I know I have opposition on this one, but I want to make a case for ONE district-wide Facebook page, especially if you have fewer than 8,000 students in your school or district. Please note that I am only talking about Facebook here. If you have Twitter and Instagram, go crazy with more pages (if you're comfortable with that). This is just about Facebook.[2]

Here are ten reasons why I always advise creating a strong district presence with one Facebook page first:

1. One page builds a **consistent district brand**. One message, one hashtag, one story, one family. Ideally, all communication from your district should have a similar message, look, logo, colors, etc. This obviously extends to your Facebook page.
2. One page builds **district pride**! The momentum builds, and once it's started, it truly is a force to watch.
3. District pride easily translates to **community engagement and community pride**. You can measure engagement on your Facebook page.

4. One message on a central page eliminates confusion and provides **consistent communication** with families and the community. Just think of the nightmare a snow day could cause if notifying parents means posting to multiple pages.
5. The Facebook algorithm – it's real! **If you have multiple pages, they compete against each other.** Rather than posting only once per day on each school page, why not combine those and get three to five posts every day?
6. **Many parents have children in multiple schools.** They want to stay up to date on all events and happenings, not just those at one school. Even worse, if one school does a "better" job of posting on Facebook than another, it creates a negative feeling toward the other school.
7. These parents and families "age up." A district-wide page ensures that **families remain followers throughout their years in your district.** If I had a kindergarten child, why wouldn't I want to see the amazing opportunities available in high school?
8. **Merging existing pages into one district-wide page is possible,** but you may need to close down pages that you no longer wish to keep. Don't worry. You can direct followers to the new page, where they will receive even more great updates about your district.
9. One district page with a few people posting **takes the pressure off schools, secretaries, tech support people, and administrators**.
10. Telling your district's story in a consistent and positive manner leads to **teacher and staff buy-in**. District pride among staff pays off big time! It builds morale, grows trust, and improves a possible negative work atmosphere.

If you are a school district with more than 8,000 students, then I would consider adding separate Facebook pages for each school. I say "consider" because you have to have someone at each school that is willing to manage the page. If you have a page, but the school is only posting once a month or even once a week, then it really isn't worth it. You also must be ready to invest in professional development for these people at each school who are responsible for the pages. Social media changes so often that it can be hard to keep up.

Even if you decide to have separate pages, you should still maintain a strong district presence on Facebook.

Staff-Run Pages

Do you have staff members who want to use social media to celebrate their classroom, organization, or team? Awesome! You want more storytellers in your district!

But you also want to provide some guidance to them. This topic comes up all the time, so I'm going to share a few dos and don'ts that you can share with your staff. This is part of a list I adapted from Mary Beddell's Communications Handbook from Plain Local Schools in Ohio.[3] We'll cover more of the specifics on what to post in the Storytelling section of this book.

Encourage staff to:

- ✔ Share photographs and videos of students and staff! Parents LOVE to see what their kids are up to and support not only their teachers but coaches, cafeteria workers, bus drivers, maintenance staff, and more.
- ✔ Help promote fundraisers and special events.
- ✔ Keep the logo consistent. However, it can be confusing if numerous Facebook or Twitter pages all use the same logo. We'll cover this more in the logo and branding section in chapter 5.
- ✔ Make the cover photo fun and change it often!

Discourage staff from:

- ✔ Using Google images, clipart, or other non-brand-specific or copyrighted photos. You have an entire school full of people to take pictures of. Unless you created it, do not use it!
- ✔ ONLY posting fundraisers or special events. You have to post "important stuff" while mixing in the "fun stuff."
- ✔ Using a bad version of the logo. No blurry logos! We want the best version of our brand available to the public.
- ✔ Making the cover a photo that is hard to see! Choose photos with lots of smiling faces.

MARY CREATED A LIST OF DOS AND DON'TS FOR HER STAFF

"I find that in my district, my staff was really open to being a part of sharing our school's story on social media, but they struggled with right vs. wrong, and they did not want to disappoint me. So much of social media is not black or white but shades of gray, so I tried to put together a list for them that was easy to understand and tangible."

—**MARY BEDDELL**, Public Relations Director, Plain Local Schools in Ohio

TRAIN AND SUPPORT YOUR SCHOOL-LEVEL SOCIAL MEDIA MANAGERS

If you have social media managers at each school, you'll need to invest in programs to train and support them. Establishing minimum expectations – for how the pages are set up, profile and cover images, and posting frequency, for example – will help. Dealing with the constant changes that the individual platforms make is another issue.

I would highly suggest having your team read this book! The chapters on branding and storytelling will provide best practices to help in their everyday management of social media.

Most school communicators train their team at least once at the start of the school year. Many try to meet monthly or bi-monthly to cover changes or updates, along with fresh ideas for content. It would be helpful to set up a Facebook group or other intranet support system where your team can share questions, concerns, and wins. Our team at #SocialSchool4EDU utilizes a private Facebook group like this for our group of social media account managers. This allows everyone to share their questions and help other teammates who may be struggling with something.

If you lack the bandwidth to provide new training topics for your team, you should consider joining our membership group. We provide weekly sessions that are broadcast live, but we record them, as well. Member schools can share those recorded sessions with anyone in their school or district. This makes professional development easy, knowing you can lean on resources that we create that will help your school stay on top of changing trends and best practices.

Which Channels Should We Use?

If you're already running social media, you may be ready for a bunch of fresh ideas. I'll give you more of those later in the book. For now, let's focus on the specifics of the channels you'll manage and the features they offer. [4]

You only have so many hours in your day. And social media is just one of your many responsibilities! So, ask yourself these questions:

- Where is my audience spending time?
- Which platform already has the most followers?

Facebook is usually the most effective channel for schools. It's a spot where you can reach parents and community members. I see Instagram as the place to connect with students and young parents. These two channels are the ones I typically recommend for all districts.

Twitter may or may not be big in your area. I have been managing a Twitter account for New Auburn School District since 2014, and I have just over 200 followers. If I needed to remove something from my plate, it would be Twitter. Of course, if your school has a vibrant Twitter presence, show up!

LinkedIn, TikTok, Pinterest, Snapchat, YouTube . . . the list could go on and on. These might be nice to have, but do you really have time to do them well?

If you ever make a decision to stop updating a platform, you could pin a post to the top of the page that states, "This account is no longer active. For exciting updates about our school, please visit XXX." There you would put the direct URL to the social channels you are maintaining.

You only have so many hours in your day. And social media is just one of your many responsibilities!

How Often Should We Post?

You need to tell a consistent story. While school is in session, you should share a minimum of two posts per weekday on your primary social channel. For most schools, that is Facebook. When school isn't in session, like summer or winter break, then you should strive for three posts per week.

For most of the schools we serve, we average three posts per day on Facebook. We pull some of those posts over to share on Instagram and Twitter as well, but not all of them. We aim to post at least twice per week on those other channels, but it really will depend on your community.

Does this stress you out? I don't think it should! There are dozens of amazing things happening in your school at any given moment. I'll give you a ton of examples in the Storytelling section so that you never run out of things you can share.

You can post on the weekend, but we see less engagement on Saturdays and Sundays because our followers are busy. You also deserve a break from monitoring new comments that may show up on these posts.

As far as when to post, we suggest posting at 6:00 a.m. and 7:00 p.m. You can catch followers when they wake up and when they start winding down in the evening. If you have more great content, other excellent times are lunchtime and pick-up time. If school gets out at 3:00 p.m., you can count on parents lining up at 2:30 to collect their little cherubs after school. What are they doing as they wait? Scrolling social media, of course! Posting at 2:30 would be perfect for a 3:00 p.m. dismissal.

You can also review your existing posting times to see what the engagement looks like. I often see social media managers posting while they are at school, at times like 9:00 a.m. or 1:30 p.m., and I typically don't see a large number of likes, comments, and shares on those posts. Just because you are posting at 6:00 a.m. or 7:00 p.m. does not mean that you have to be online at that time. It simply means you will schedule your posts to be shared at that time. Putting yourself on a typical posting schedule will help you stay organized and will also get your followers used to a schedule.

My advice to post this often will definitely contradict business social media training that will tell you to only post once per day or even less. I've been doing social media since 2014, and I've worked with hundreds of schools directly. Schools are very different than businesses. You have an engaged audience that wants to see what is happening in your buildings. Posting only once per day or every few days would not give you the chance to celebrate everything that your students and staff are doing.

Social Media Presence Driven by Data

In a podcast interview for *Mastering Social Media for Schools*,[5] I spoke with Angela Brown, a K–12 senior enrollment insights leader for Niche. This platform provides in-depth profiles on every school and college in America, offering reviews and ratings, search and data tools, and recruitment information based on surveys. In our interview, Angela Brown shared insights on how certain social media platforms can drive interest for enrollment. Here's a little of the gold mine of info she shared:

- ✔ **Facebook** – Of parents surveyed by Niche, 40 percent said Facebook impacted their enrollment decisions.
- ✔ **Instagram** – The Niche polls showed Instagram influenced 20 percent of parents.
- ✔ **YouTube** – The data from Niche showed YouTube, which is a Google property, to be useful in terms of recruiting new students and getting parent attention. "When we looked at our top social channels, YouTube was in the top three, both with students and parents," Brown reported. Nine percent of parents said it influenced their decisions.

TikTok didn't make the top three with parents, but the short-form video for Reels can be cross-posted without investing a lot of extra work.

Here's why it's worth paying attention to social media data: Niche surveys showed that 60 percent of parents said their children played a role in school choice. When narrowing the survey to parents of high school students, 76 percent said their children influenced their decision. Angela said it's "something to think about as you're planning your social strategy and having a dual strategy for how you're going to engage the adults, but also how you're going to engage the students. And you can do that with different types of content because a lot of the platforms are the same."

Which platforms influence the students?

1. Instagram – 25 percent
2. Facebook –12 percent
3. YouTube – 11 percent
4. TikTok – close to 11 percent

Data from sources such as Niche can help you see how specific social media platforms might benefit your school. A word of caution, though. Just because you have data doesn't mean you can or should go all in on *every* platform.

Grab It!

Listen to the podcast interview with Angela Brown titled **"What the Latest Data Reveals About Parent Behavior and School Marketing Priorities"** on *Mastering Social Media for Schools*. Get the links to the enrollment insights data she mentioned from Niche in the bonus resources. The K–12 research can help you make informed decisions about your social media presence and website content.

Social Media Is Like a Free Puppy

What might seem like good intentions – communication and celebration – can quickly turn into inactive accounts, problematic comments, and total chaos around what parents should be following and how they should stay updated.

Now, I am not one to try to muffle anyone from celebrating the great things happening in your school or district. The more people telling your story, the better.

But . . .

You and I both know that managing social media pages is not just about posting updates. You actually have to be ready to respond to questions, deal with frustrated parents, and realize that the algorithm won't show content to everyone who follows the page.

In other words, as my dear friend Kristin Magette, APR, explains, "Social media is like a free puppy, not a free beer!"[6] The beer comes without strings attached or a long-term commitment. But a puppy – that's a commitment. There are a few hurdles to consider when considering adding new channels.

Hurdle #1:

Each separate channel means building a new audience or following.

Imagine going to the circus. There is a big show under the main tent that everyone flocks to see. That is like your district's social media accounts: Your primary Facebook, Twitter, and Instagram accounts.

Now, at the circus, you can also travel from sideshow to sideshow to watch specialized tricks or acts. But circus attendees need to walk around to see those – and if they aren't careful, they might miss the big show!

These sideshows are like all the other social media accounts associated with your school. As a parent of six kids, I can say firsthand that by trying to follow a district account, a high school page, a middle school page, an elementary page, a basketball page, an FFA page, a band page, and more, I am bound to miss something important!

Hurdle #2:

Every post does not reach every follower. Many social media page managers don't realize that posts made on your page are not seen by all followers. This is due to the algorithm that exists on each platform.

There is so much content out on social media that Facebook and other platforms can't show every post from every friend, business, or page that a Facebook user follows. They have an intricate algorithm that determines who will see what posts. This involves how often the account posts, whether people are engaging with it, and more.

Your district channels likely have the biggest following, so celebrating the big milestones and accomplishments for classrooms, organizations, and teams on the district's social media accounts makes the most sense.

If your staff members need a way to communicate with the people most related to their program, I suggest direct email, text, or other apps like Remind, Class Dojo, or SeeSaw. Those methods are more apt to reach the intended audience.

Hurdle #3:

Are you ready to respond to strange comments? Posting on social media is just half of the equation. Social media is meant to be social – meaning people will be interacting on that new account.

Are your colleagues prepared to monitor the account for questions that pop up from parents who expect an immediate answer from you? If someone starts complaining or arguing with another follower of the account, are they prepared to handle it?

This is the part where the free puppy gets sick, and you have to take time out of your busy schedule to take it to the vet.

If one of your staff members is going to start their own channel, they have to be prepared to deal with comments. And it's not as easy as turning off comments altogether.

Hurdle #4:

What is your district plan for these account requests? Developing guidelines to handle these types of requests for social media accounts is important. Will you allow Twitter accounts but not Facebook pages? Can only employees of the district-run pages be associated with the school and not volunteer coaches or parents? And what happens when someone leaves and doesn't shut down their account?

If you are ready to tackle this – and it is not for the faint of heart – please know that it will likely take years to get it under control. If you allow pages from others, you should consider an official social media directory on your website. This helps parents and community members find the correct accounts for your school.

Lana Snodgras, the director of communications for West Plains School District in Missouri and whose case study I shared in chapter 1, has a social media account request form that faculty and staff need to fill out if they would like to set up a special page. That form includes details required for approval:

- ✔ The account manager's name
- ✔ The name of the proposed page
- ✔ The target audience
- ✔ Proposed content
- ✔ Plans for monitoring the page
- ✔ Policies and guidelines about content
- ✔ A signature of building administrator approval[7]

Understand this, too: not everyone will ask you before they start the account. Some of your colleagues will just start them, and you'll find out about them later. This will be an ongoing process and requires staff training and constant communication!

What if you've already established channels you can no longer manage? It might just be time to shut some of these other pages down or simply state that the page is not currently being updated. Direct them to a social channel that you actively manage. And then, feel great about your decision!

Do I "Reely" Need to Do This?

One question that may have come to mind when I listed the platforms above is: Do Reels matter? We'll be covering specifics of content later. But as you think about platforms, it's helpful to consider what content will have an impact and where. This is always changing, which means sometimes you'll need to change up your plan based on what is most effective. Here's an example from my experience.

I started adding Instagram Reels to my school page for New Auburn in the fall of 2021 and saw my reach on Instagram grow by ten times. The dramatic reach has calmed down a bit with algorithm changes, but I still think you "reely" need to add Reels to your strategy. And for me, adding Reels was much easier than starting a TikTok channel.

There are many ways to use this short-form video option within Instagram and Facebook. You can create a slideshow with photos, combine photos and videos, or simply upload videos from your school. You'll be able to bring them to life with stickers, text, and music.

Grab It!

Instagram Reels is a feature within the Instagram app where you can create short videos. It might sound intimidating, but I promise you – if I can figure it out, so can you! I have a super helpful blog linked in the free bonus resources for this book. Grab **"How to Use Instagram Reels for Your School"** at socialschool4edu.com/book.

That bonus includes:

- How to get started with Instagram Reels for your school
- How to use Reels
- Tips, tricks, and hacks you must know
- Examples from K–12 schools
- Creative ideas

As you engage your community with your school, I encourage you to start with what you can manage right now. If that's a Facebook page, begin there. As you grow and expand, I'm here for you with hundreds of helpful resources in this book, on my blog, and in my membership group. I believe in you!

YOU'VE GOT THIS!

- [ ] Establish one district-wide Facebook page.
- [ ] Provide staff with clear guidelines for staff-run pages.
- [ ] Ask: Where is my audience spending time? Which platform already has the most followers?
- [ ] Think carefully before you adopt a new "puppy," aka social media channel.
- [ ] Give yourself permission to shut down channels that aren't serving you well.
- [ ] Maximize the platforms you're on. For example, use Reels on Instagram and Facebook.

BRANDING

BRANDING

LOGOS, COLORS, AND FONTS

5

Case Study

Meet Megan Anthony: Megan is the communications coordinator for Canal Winchester Schools in Ohio and, in her previous role, helped serve four different Ohio schools and her educational service center with communication services. She worked with Greenon Local School District in Ohio on a rebrand. Megan is an active part of the #SocialSchools4EDU membership program.

The challenge: There was very little consistency around the use of the school logo. The athletic department had started using a completely different logo that the community loved, but it wasn't approved by the superintendent or the board! Brand confusion and frustration ensued.

The process: Megan's rebranding effort came after a $50 million bond issue passed, and the district was building a brand-new K–12 school. She saw her opportunity and began talking to district leadership. "We have this brand-new school! We should do the work to get a strong visual brand so that we have the right branding throughout this school. We can't invest $50 million into something that isn't really showing who we are!"[1] she said.

Megan built a committee of key stakeholders who were committed to seeing the rebrand project succeed. She also looked for professional consultants with experience in school rebranding to help guide the rebranding process and identified other tools (like community surveys) to make the project

If you don't have a consistent logo, colors, or style, your school needs branding help. But you don't have to do it alone. There are professionals to help you with the entire process – and it will pay off big when you and your community are proud of your brand.

CASE STUDY

happen. She had some pushback but handled it professionally. "We really had to confront some attitudes and feelings to make sure that it did not end up being just one group of people who were guiding the process. Make sure you have an idea of what is it that made them rally around this. What's important to them? And then communicate the value of the branding process. Tell people what you're doing and why you are doing this, and why it's important."

Using Google forms, Megan conducted in-depth surveys to gather valuable input and allow people to feel heard. Then, she conducted smaller focus groups. She worked with a professional designer, kept the community involved, and was prepared for any of the final logo choices to be the winning design.

The outcome: The rebranding process took one full school year; it kicked off in September, and the final logo was approved in June. They ended up with a logo and design the community could be proud of! You'll find a link in the bonus resources to the Greenon Schools official brand style guide, but here's a peek at one variation of the finished logo. Isn't it amazing?

A brand is more than a logo and school colors. A brand is an identity, a voice, an unambiguous presence that is identifiable with you and only you as a school district.

But in my many years working in school social media, I have seen the good, the bad, and the ugly when it comes to school branding! We are all in the space of continuous improvement, so this chapter will serve as your roadmap to leading a rebrand for your school – whether you're currently going through the rebranding process or if you see it as inevitable in the near future.

And if you're branding a new school from scratch – awesome! This chapter can still help you establish the needed brand guidelines to set you up for success. As I said above, a brand is more than a logo and school colors, but for the sake of this book and how it pertains to social media, we are going to focus most of our attention on that.

Do You Have a Logo Problem?

Even though a brand is more than a logo, it often starts there. A logo is a graphic mark or emblem that is used to promote instant public recognition of your school. It is the visual representation of your brand. In a live training session for our #SocialSchool4EDU membership program, Megan Anthony, Communications Coordinator for Canal Winchester Schools in Ohio, shared a list of signs your school might need a rebrand. Let's look at a few.

- ✔ You have several versions of a logo that look nothing alike in coloring or style.
- ✔ The logo is a recolored version of a pro team logo, which has potential copyright issues.
- ✔ You've updated your school mission statement or core values.

- ✔ You have a new school mascot.
- ✔ You've restructured your schools or attendance zones.
- ✔ You passed a major bond/referendum.
- ✔ You have not revisited the brand in more than ten years.[2]

It isn't easy to lead a rebranding effort. Change is hard for many people, and alumni and community members will have an emotional attachment to the current brand. You could even keep the same mascot, and people will get emotional!

Rebranding your district can feel a lot like having a root canal without any anesthesia, so do yourself a favor and realize that projects like these take time. It will take determination and leadership to get it done.

A logo is a graphic mark or emblem that is used to promote instant public recognition of your school. It is the visual representation of your brand.

Leading a Rebranding Effort

As you lay the groundwork for a rebrand, you'll want to ask the three questions Megan Anthony used before she began gathering the data:

1. **What's wrong with the current branding?** Identify the problems – such as a lack of consistency, lack of knowledge, or general brand confusion.
2. **What's right about the current branding?** What do people love? Where are their emotional ties?
3. **Why now?** What makes this the right time to take on this initiative? Why should it be a priority? Are other changes coming up (new athletic uniforms, a new school address, etc.) where it just makes sense to rebrand now?[3]

Once you've built a strong "why," it's time to make your pitch to the people in charge of approving your effort. Megan advised building your pitch around the following questions:

- Who is your audience?
- Who are your allies?
- Who do you need to convince?
- What are the pros for rebranding?
- What are the cons?
- What resources do you need? [4]

As you build a committee of key stakeholders who are committed to a successful rebrand, you'll also need professional consultants with experience in school rebranding to guide the process.

When it comes time to engage people in the community, it can get messy. School leaders, teachers, parents, community members, current students, future students, coaches, and alumni might all want a say in the process, and some will be part of the process. No matter who has a seat at the table, your main goals remain the same.

Megan shared three goals to keep in mind:

Goal #1: Identify your audiences and make sure they are heard and represented.

Goal #2: Understand what your audiences love about the current branding.

Goal #3: Communicate the value of the branding process. [5]

Throughout the process of listening and handling pushback, those goals will remain front and center. When people understand the value of a process, they are more likely to buy into the enthusiasm. Their feedback can be a double-edged sword! But in the world of education, it's important that we are transparent and focused on getting buy-in.

Some tips to help in the design process:

✔ Work with a professional graphic designer.

✔ Narrow ideas down to a few options.

- ✔ Survey the audience to pick a winning logo, including students of all ages in the voting. Kindergartners can circle a favorite logo on a sheet of paper!
- ✔ Take your time in the process. Don't rush it. A rebranding process might take a full school year.

A BRANDING WIN!

In a podcast interview on *Mastering Social Media for Schools*, Ian Halperin shared with me how a branding campaign was one of his school's biggest social media wins. His district has multiple campuses, including two high schools. Each campus has a mascot logo, but they developed one district logo in black and white and then created color variations for each campus. That logo was a simple W in a circle with the district name at the top of the circle and space to have each school campus name on the bottom.

Ian said, "It grew on me." He said his first thought was, "It's just a simple W. But we did car stickers with every school campus. We did little flags . . . and now you drive around town, and you see all the different school colors." He sees them all over Dallas, he said. Campus web admins pushed out the style guides throughout the district. "So now we have uniformity across all of our campuses." Having the superintendent behind the efforts made all the difference, Ian said.[6]

—**IAN HALPERIN** is Executive Director of Community Relations and Marketing for Wylie Independent School District in Texas.

You'll find a link to the Wylie ISD style guide in the free bonus resources at socialschool4edu.com/book.

Roll Out the New Brand

In a guest blog for #SocialSchool4EDU, Heidi Feller shared elements of a "Roadmap to Rebranding" presentation that Megan Anthony created for our membership program. The following is not meant to be a comprehensive list, but it should help you cover your bases as you move through this process. Heidi said, "Remember, it is a process! It will take time for your staff and community to adjust."[7]

Make your new logo, colors, and brand a BIG deal. Splash it everywhere – and get swag!

You could create a bumper sticker for district families. Make a special announcement on social media. Consider a video with the superintendent. You could even give away staff T-shirts proudly featuring your new logo.

Ask people to send you pictures of themselves wearing the new swag for you to post on social media. You could also ask staff members to wear it each Friday. Trust me, giving staff a free shirt will lead to more people being excited about your new brand!

It might take a while before your school has the budget to purchase new athletic uniforms to implement the new brand. It's OK if the old branding is "in rotation" for a while longer. [8]

BRANDED SOCIAL MEDIA TIP

Look for ways to customize your theme color or fonts on any platform you use. The color codes in your style guide can be added wherever a program allows you to customize.

For example, because you can change the background colors of your Instagram Stories, it's also possible to use your school brand colors. You can also create branded templates in your favorite design program – I like Canva.

Grab a link to a full tutorial and more customizing help in a blog post titled **"Our Very Best Instagram Stories Tips for Schools!"** That's in the free bonus resources at socialschool4edu.com/book.

School or District Name

We've spent a lot of time here talking about logos, but there's one more part of your brand that matters: What is the official name of your school?

For example, is it School District of New Auburn or New Auburn School District? Your district may have a number associated with it, so should you be calling yourself ISD 803 or Wheaton Area School District? Is "district" part of the official title? If so, it should always be capitalized with the name. Is it OK to drop "school" from the name in social media posts? For example, referring to Parkview Elementary School as Parkview Elementary.

Decide on the official title to use for all communication. If you have an approved acronym, always use the full name in the first reference with the acronym in parentheses before you use the acronym by itself. Example: New Auburn School District (NASD) is a pre-K through 12th grade public school.

Be consistent. Make sure everyone knows how the school or district name should appear in email signatures, letters, and public posts.

If your district goes through a name change, it can take a while to get everyone on board with using the correct title. It helps to explain why it matters to use the new name in all communications. Reasons include consistent messaging, clear communications, and presenting a united image.

Once you have your brand elements and school title established, it's time to make sure everyone knows how to use them and create a plan to maintain them.

Create a Style Guide

A style guide contains specifics such as color codes, fonts, and logo forms, Heidi Feller said in a guest post for #SocialSchool4EDU.[9] Place the style guide "on your website and direct inquiries to that page," she said.

The style guide also includes some of the information we just covered about your school's name. Give that guide to everyone who creates anything that visually represents your school: email signatures, classroom newsletters, tickets to events, shirts – ALL materials.

Your style guide will include details like these:

- Instructions that the logo must be used as a complete entity – no extracting or cropping.
- Instructions for not stretching or recoloring the logo.
- Visual representations of the main logo and any secondary logos.
- Color swatch codes.
- Official district font names (also called typefaces) and visual examples.
- How to use the logo.
- Visual representations of good and bad uses of the logo.
- A list of templates and tools the school has already created for use.

You can also show an example of an email signature with the school logo so staff members have an easy template:

Staff Name
Title
School Name

Phone Number
Email

School Website
School Hashtag
Logo

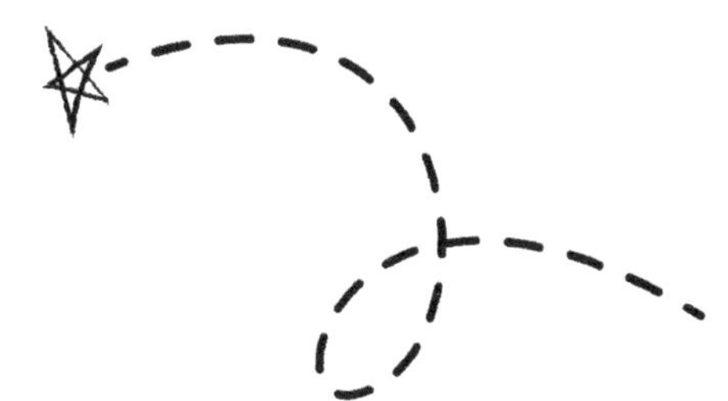

Grab It!

There's a link to examples of **logo guidelines and a style guide** from the Greenon Schools and South Washington County Schools (SoWashCo Schools) in the free bonus resources. Note how SoWashCo, a suburban district located southeast of St. Paul, Minnesota, also incorporates approved school titles/names in its brand guide.

Brand Maintenance

Brand maintenance is an ongoing process that includes gentle yet firm reminders about the correct brand usage, Heidi Feller explained in a blog post for #SocialSchool4EDU. It's never-ending but well worth it to protect your brand integrity.[10]

Megan Anthony gave an example from Greenon Schools in Ohio. She suggested identifying **brand managers** in your district who are empowered to enforce brand standards and grant logo usage requests. She also recommended having **brand users** who can order school swag and branded items, as well as **brand partners**, which are external groups (PTOs, local businesses, etc.) who have permission to use your logo.

Megan said, "We invited all of those people to a meeting where we shared the brand guidelines and answered any questions they had. Everybody got my email address, a copy of the brand guidelines, and an email with all of the logo files."[11]

Brand maintenance is an ongoing process that includes gentle yet firm reminders about the correct brand usage. Heidi Feller

---

Part of your brand maintenance will include scouting around the school and watching communications for old logos that need to be replaced. For example, take a look at the letterhead and envelopes used in the school offices. Help secretaries understand what the new branding means for them. Ask them to keep an eye out for old items and to direct people to the brand guidelines as needed.

Send the logo to staff members for use in emails and on their classroom websites. Sending JUST the logo in an email, along with instructions on updating a website logo, goes a long way! Don't assume that people know how to switch out a logo, so recording a short video to include in this correspondence will garner extra friends.

Don't forget about food service and transportation departments and keeping your athletic director in the know.

YOU'VE GOT THIS!

- ❑ Review the logo(s) in use in your district for consistency. If there are several unrelated versions, it's time to rebrand.
- ❑ Work with professionals to establish a logo, colors, and fonts for your school.
- ❑ Involve the community in the process by gathering input on surveys and letting them vote on the final logo.
- ❑ Celebrate the big reveal with new swag!
- ❑ Create a style guide and make it accessible to everyone.
- ❑ Make sure every department has the logo and uses it.
- ❑ Maintain the brand assets diligently and gently but firmly correct when you see misuses or straying. Toss the outdated stuff.
- ❑ Be patient with athletic budgets and waiting for new uniforms.

YOUR BRAND VOICE

6

How would you describe your school in three words if it were a person? What messaging and tone make up your school's voice? You'll be ready to develop a profile for your style by the end of this chapter.

Case Study

Meet Marissa Weidenfeller: Marissa is the former communications and community engagement specialist for the School District of Fort Atkinson, Wisconsin. The school district has 3000 students in four elementary schools, one middle school, and one high school. She has a marketing and PR background. Marissa is all about strong branding and making sure families have a consistent experience.

The challenge: Marissa had a wonderful group of people assisting her with posts from the different locations. They definitely knew their schools, their cultures, and the brands of each. However, Marissa needed to bridge a gap between the brand of the schools and the brand at the district level to make sure it was cohesive.

The process: Marissa and her team came up with an initiative they called "1FORT," inspired by their town of Fort Atkinson, to demonstrate that the businesses and the schools were all united. They created bumper stickers, yard signs, window clings, and more, all displaying the 1FORT theme.

The outcome: What was at first a small initiative became a voice for the whole community. The mantra has continued beyond the initial campaign. While working in Fort Atkinson, Marissa said, "We're one team, one district, and one community, and that's our brand. And our community knows it." That got them through a pandemic, but the phrase has

CASE STUDY

become the voice for what gets people through all sorts of challenges. It directed communication in a positive direction through the pandemic and united people. The school "voice" ultimately became a community mindset.[1]

What action steps have you taken so far?

Tag me on Twitter at @andreagribble and let me know!

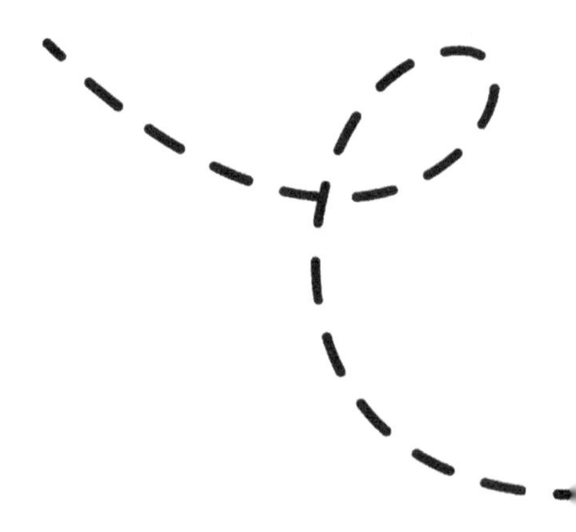

What do you *say* in a post? Have you ever found yourself in a creative rut when writing captions? Do your posts feel as if the words have a sparkle of life, or are they matter-of-fact lists of information? It isn't uncommon to run out of ideas to make a picture sound exciting.

When you find your social media voice, then you can zero in on that while using creative writing in your captions and calls to action. And you can practice being consistent in your tone, even if you have multiple people on your team creating posts.

What Is Voice?

Your brand voice is how your followers perceive your school's personality by the tone you use in communications. It has to do with how your communication makes the audience feel, too. Social media captions can convey authority, playfulness, a casual tone, knowledge, confidence, and more. You might notice that one teacher or school administrator uses a different tone than other teachers and staff. One might be formal and the other casual.

If your logo didn't appear with your content, would your followers know it's your school as they scroll through your posts? The colors and style in your brand graphics and photos can also contribute to the tone of your voice. Dark visuals with bold phrasing convey something different from lighter, more casual images. You can have the same consistency with your captions as you can with your graphics.

Social media posts are not press releases.

Some businesses have built social media reputations around being sarcastic, irreverent, and even self-deprecating – styles that might not have the same effect for an educational organization. However, your school doesn't have to sound formal and buttoned up, and your posts can be creative and witty in their own way. Social media posts are not press releases.

As you develop a voice, think about who your audience is and what speaks directly to them. You can convey qualities such as confidence, reliability, and an ability to make learning fun in how you share information. When you use a clever play on words, for example, you're conveying fun alongside the message of learning. You can also communicate enthusiasm and school spirit with the right combination of words, punctuation, emojis, and your hashtag.

Your content conveys a mood through the style of the words used, the way you use language, and the mood it expresses. Let's look at how to identify the voice of your district or school.

You can convey qualities such as confidence, reliability, and an ability to make learning fun in how you share information.

Captions That Fit Your Voice

You know that you can't just post photos. And then there's the important goal of driving likes, comments, and shares. We'll talk about hashtags (chapter 7) and opening questions later (chapter 11). And I'll give you plenty of caption ideas, too (chapter 14). But where do you find ideas for writing more creatively without losing your unique voice?

At one annual retreat, two of our amazing leaders at #SocialSchool4EDU shared a great breakout session on creative writing. I think you'll appreciate their simple strategies and tools![2]

Assume that every story submission you receive will need finessing. We receive photos and short stories from staff members at the schools we partner with. It's not enough to just post what they provide! We always modify it, and here are some ideas:

- ✔ We look for ways to include the hashtag.
- ✔ Look at other districts if you need inspiration or are in a rut. You can follow #SocialSchool4EDU on Facebook for some ideas!
- ✔ Use a short phrase or headline at the beginning, then dig into the facts.
- ✔ Screenshot ideas when you're scrolling on social media! You can save these screenshots into a folder on your phone. You can also use the save feature in Facebook and Instagram. Following other school pages is great, but make sure when there is a post that catches your eye, you save it!
- ✔ You can also get great inspiration from business and non-profit social media pages. Save those ideas, too!
- ✔ Use Pinterest to search for great quotes. Simply type in a keyword (like "smile") along with the word "quote." You'll find great options!

If you're looking for the best creative words and phrases to start a post, these are some we use often:

- Check this out!
- Who do you recognize?
- #TBT Allstars
- Throwing it back to the late 90s
- Caught "read" handed
- Help us caption this photo!
- Extra, extra, read all about it!
- Our students are awesome! Wouldn't you agree?

When you finesse text, always remember the purpose of telling the wonderful stories of your students. Why make it matter-of-fact when you can convey excitement and enthusiasm for the great things happening in your school? I've included examples in the sidebar of how we modify text to fit our brand voice and social media goals.

Quick tip: If you're searching for an example on a specific topic on another school's Facebook page, try this! From your computer, go to the Facebook page and then use the "search" function. Any posts with that keyword will appear. Look for creative ways they captioned posts to screenshot for ideas.

SCAN QR CODE

FOR BONUS RESOURCES

socialschool4edu.com/book

Grab It!

As an educator, you know you need to communicate with your families, but it's hard to know what to say and how to say it. After a fifteen-year career in journalism, Patricia Weinzapfel, MS, a school home communication consultant, found herself in a room full of educators and discovered she didn't understand the acronyms and words they were using. So how could families understand it? In podcast episode 113 of *Mastering Social Media for Schools*, Patricia shared simple tips to instantly improve your writing and tone. Grab the link to the episode, "**Building Strong Parent Partnerships with the Power of Our Words,**" in the bonus resources and get tips such as:

- Don't use a five-dollar word when a fifty-cent word will do.
- Use the word "you."
- Check your tone by reading your words out loud.[3]

EXAMPLES OF HOW TO FINESSE TEXT

Before (the text that came with the photo submission): Students in the yearbook class chose to come to the library to help with moving boxes (the library has to be boxed up for renovations this summer). Their help was very much appreciated.

After (how we phrased it for posting): "Many hands make light work!" Students in the Chi-Hi yearbook class helped pack boxes in the library, which will undergo summer renovations. Thank you! #MightyCardinals

Before (subject line of email): Homecoming Day 1 – USA Day.

After (the post): Stars and stripes forever! ??[heart emoji]? High school students went all out for their first homecoming dress-up theme today . . . "USA Day." Hooray for Homecoming 2019! #CSDGoBeavers

Before (subject line of email): Whatever it takes.

After (the post): "Our #NewAuburn teachers do whatever it takes to serve our students! We've had a great start to our school year – and teachers and support staff – we see you. Thank you for your dedication to our kids! [emoji]

Ready to try it? Rewrite several of your captions to add some creative flair that reflects your voice and engages your followers!

Stop using five-dollar words when fifty-cent words will do. For example, don't say "inclement weather" on your calls home to cancel or delay school. "Due to the poor weather conditions, school is canceled." Isn't that easy?

Be Memorable, Plus Create with Intention

Every social media post has the chance to tell a story. Are you creating with intention? When you create a caption with intention, you're putting thought and strategy behind it. School communicator Kristin Boyd Edwards believes in two goals for content creation.

One goal could be to "stop the scroll." In other words, you want to catch people's attention as they mindlessly scroll through their social media feed with a story that is "sticky," or memorable. You want your school or district to stand out – but more than just one single post. You want your voice to be something that's cultivated and memorable, solidifying your brand in the minds of your audience far and wide.

Another goal could be to generate engagement. You want people to like, comment, view, and share your social media posts and videos. Engaging content is more fun . . . and memorable![4]

Taglines and Slogans

A unique school tagline or slogan isn't just a catchy phrase. It represents the mission and vision of your school. A tagline is part of your brand, but not every school needs one. Sometimes a catchy phrase can work well with your brand to help convey your mission. Other times it isn't necessary because you have a hashtag that says it all.

Tagline examples:

A great place to grow and learn – Grand Forks Public Schools, North Dakota

Guiding Students. Empowering Futures. – Beaver Dam Unified School District, Wisconsin

Committed to Academic and Personal Excellence – Fall Creek School District, Wisconsin

Together we thrive – East Jackson Community Schools, Michigan

Engage. Equip. Empower. – School District of Altoona, Wisconsin

As you think about your brand voice, this is a good time to evaluate your tagline, if you have one, to see if it communicates what you want people to "hear" about your school. It's also a great time to see how it fits with your hashtag.

Grab It!

See how South Washington County Schools (SoWashCo Schools) incorporates messaging and voice in its style guide. The link is in the bonus resources at socialschool4edu.com/book.

Develop a Style Guide for Your Voice

Whether you have multiple contributors for content or you're the solo communicator, including brand voice in your style guide along with the graphics, logos, and fonts can help you develop consistency. It can also help to make it easier to incorporate new team members when transitions occur.

Here are a few tips for creating a simple style guide.

- Review your school mission and values.
- Include a list of who you are and who you are not. Examples: Our brand voice is confident and celebratory. Our brand voice is not snobbish or too casual. Our brand is helpful and authentic. Our brand voice is not formal and does not use big words people don't understand.
- Write a paragraph that describes a simple profile of who you serve. Then imagine writing captions and communicating directly to that person.
- Look at your best-performing posts. These resonated with your audience the most. What tone do you think the captions conveyed?
- List three to five key actions that are part of that brand voice. Examples: We use puns and plays on words to have fun. We

always use the school hashtag to grow school spirit. We strive to be helpful, so we provide clear action steps for our followers.

- See if you can describe your brand in three words. Example: If [our school name] were a person, I would describe it as [three traits]. Then define those traits with a short description of each. Need inspiration? Ask some staff members how they would describe your school in three words.
- Revisit the brand voice periodically and look at some of your recent posts. Does your content fit the persona you established? How can you tweak it from there? Sometimes your branding needs to be adjusted, and other times your content needs tweaking.
- Create a master document that lists your brand voice attributes and your dos and don'ts for social media posts.

PODCAST BECOMES SCHOOL "VOICE"

ERIN MCCANN is the former director of digital media and marketing for Allen ISD in Texas. She and her fellow school communicators wanted a creative way to educate their Texas community on the "why" behind the need for a voter-approved school bond that would allow the school district to finance capital needs. They decided to create a podcast to catch people on their commute with information. "That was everything from an interview with the tennis coach about why the tennis courts were up for renovation to an elementary principal on what a new campus might look like. Or our technology director and what we would be doing with one-to-one devices." To comply with laws about districts not advocating for a bond, they were careful with every piece they put out. "It was all purely informative," Erin said.[5]

If you're looking for even more help on the creative writing side of social media, the rest of this book won't disappoint! I'll cover a lot of ideas for writing great captions. You want your followers to feel a strong connection with your school. So, for now, focus on how your followers can see your school as a trusted friend. Listen carefully to how people perceive your school, and continue to work on how your voice tells your story.

YOU'VE GOT THIS!

- ❑ Develop your own style guide to help all team members know your voice.
- ❑ Look for ways to finesse text to make it more interesting, to include your hashtag, and to encourage engagement.
- ❑ Screenshot and save ideas from other districts or organizations to use for inspiration.
- ❑ Remember, a brand is an overall strategy and message behind your logo, graphics, mission, tagline, *and* voice – not just a logo.
- ❑ See if you can describe your school as if it were a person, using three words. Do your posts convey this persona?

#HASHTAGS

7

Case Study

Meet Sherese Nix: Sherese is the executive director of communication for Garland ISD (GISD) in Texas, a district of over 55,000 students with seventy-two schools. She's been a teacher, an administrator, an assistant principal, a principal, and chief of communications.

The challenge: GISD needed a hashtag for the district that would empower teachers and administrators to take ownership of the stories they were telling. Sherese and her team wanted something that made an impact.

The process: They landed on the hashtag #TheGISDEffect. Everybody has an impact, Sherese said. "Whether you're a custodian, whether you're a bus driver, whether you're a food nutrition worker, whether you're a teacher, a principal, central admin – something that you do every day is going to have an impact in the child's life. So why not understand that? Recognize that and empower and use that effect." The tagline for the district is "Changing lives and impacting futures."

The outcome: The GISD Effect has become a nationwide movement, spreading beyond Garland as students have graduated, gone to college, and moved on into other areas of life. Ultimately, Sherese said, those individuals *are* the GISD effect. The concept of changing lives and impacting futures extends to universities, businesses, and beyond. Students tag themselves in their own posts, saying, "I am #TheGISDEffect."

A unique hashtag for your school will not only help everyone celebrate your school, but it will also help you find great content. You need a fun hashtag that everyone can use!

CASE STUDY

The hashtag conveys what the district is about. It's about creating an experience, shifting the frame of mind, and having a positive impact. Sherese said, "I believe in creating experiences. When you connect with people, when you provide an opportunity where people feel valued and valuable and appreciated and loved, that's when people want to come to your district."[1]

Every day, the GISD community impacts lives and changes futures, proving they are all connected to something bigger than themselves. The hashtag has become a representation of that impact.

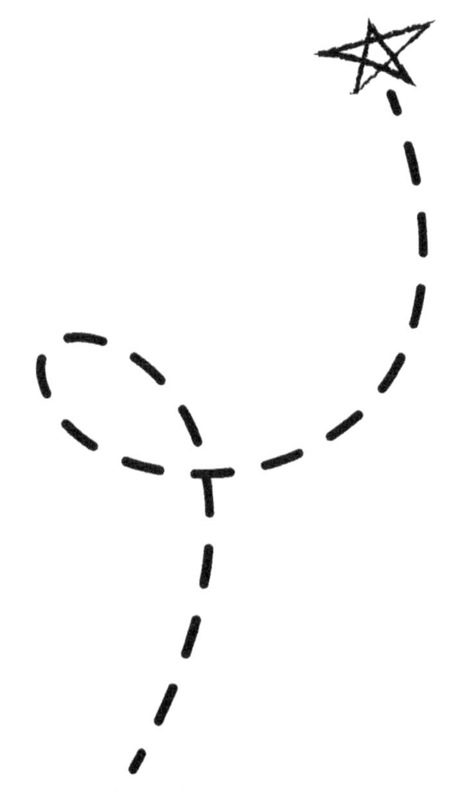

It isn't a pound sign, nor the symbol for *number*. It is now officially known as the hashtag. Children born after the year 2007 don't have any idea about the old definitions.

Understanding why and how to use hashtags is an important aspect of telling your school's story. With every presentation I do with districts across the Midwest, I get at least one question pertaining to this mysterious little symbol. So, let's break down some FAQs when it comes to hashtags.[2]

Proving my point is a story of my daughter when she was in first grade. The teacher was doing a math lesson and wrote "# of cookies" on the SMART Board. Kyra quickly raised her hand to ask, "Mrs. K, what does hashtag of cookies mean?"

What Is a Hashtag?

Simply put, a hashtag is an easy way for people to search for tweets or posts that have a common topic. In this case, that topic is your school. Hashtags consist of the # symbol followed by letters or numbers. No spaces can be used, nor symbols. Capital letters matter for people who are visually impaired and use screen readers for accessibility. However, capital letters don't affect the hashtag itself.

So, if you happen to love NCAA basketball, you could follow #NCAAhoops. That could also be typed as #ncaahoops. If you were to type in #ncaa hoops with the space, then that will not work. The hashtag recognized will be #ncaa, so you may get stories on football or gymnastics that pertain to NCAA.

You can get in the conversation by utilizing the hashtag in your post, or you can simply read other posts using that hashtag. Many users of Twitter or Instagram, where a hashtag search works the best, use hashtags to consume or find information. They may never use the hashtag themselves.

Why Do We Need a Hashtag for Our School?

Hashtags allow everyone to become a storyteller for your school. At any given moment, there are hundreds of things happening inside your district. If you want a chance for all those things to shine through, you need a hashtag that everyone uses on social media!

It's fun to find and share content that you locate by doing a hashtag search. If you repost photos on your own school page, you should always give a thank you to the originator of the content. Retweeting is another option to simply share the post on your Twitter feed. You can also easily share Instagram posts to your school's stories or use an app like Repost to put them into your feed with proper credit.

Hashtags allow everyone to become a storyteller for your school. . . . You need a hashtag that everyone uses on social media!

Is There Such a Thing as Too Many Hashtags?

Yes, there is such a thing as too many hashtags for your school. If you end up asking yourself which school hashtag you should use on a given post, then you have too many. I think it is very important to establish one primary hashtag for your district first. Promote it every chance you get so that you can get your channels flooded, like the Fall Creek School District in Wisconsin does with its use of #GoCrickets.

Once you have one that is well known, you can possibly add separate hashtags to differentiate between schools, sports, or other special interests. Additional hashtags could be used in your community, but before you start utilizing them in school district posts, you'll really want to make sure you want to promote them. An easy way to decide is to type that hashtag into the search bar of the social media platform. What kind of content comes up?

GRAB FOLLOWERS AT THE ENTRANCE

Does every person who walks into your school realize you're on social media? With our partner schools, we've added weather-resistant stickers to their entry doors with the school hashtag and the Facebook, Twitter, and Instagram logos and handles.

We provided the sign company with the school logo, hashtag, and handles, and they did the rest. The signs are about 14x14 inches, but you could go bigger if you wanted.

Please note that if you choose to promote your channels in this way, you'd better be consistently updating them! Promoting a Twitter account that hasn't been tweeted on for two years will send the wrong message to your school community.

What If Someone Uses the Hashtag in a Negative Way?

When you utilize a hashtag for your district, of course, it's a possibility that someone could use it in a negative way. A neighboring school rival could use it to bash your basketball team, or an angry parent could take to social media to gain attention on a complaint they have with the school.

Social media provides a stage for positive and negative things for others to say about your district. However, the more people use your hashtag to celebrate all the positives about the school, the less likely it is that people are going to take notice of the negative posts.

Please don't let the possibility of something negative showing up with your school hashtag deter you from promoting all the awesome stuff that is happening. People will say not-so-great things about your district with or without your school hashtag. Consider it a bonus that you can actually see it![3]

A HASHTAG SEEN ALL OVER THE COMMUNITY

We are the Hurricanes, and we have a saying that has been used for years – "like a cane." It means doing your best every day. Around our community, we see more than 150 banners with "Like a Cane" on them. Businesses and organizations have joined in the campaign to do their best. Now the campaign has transformed into our district hashtag #LikeACane! We incorporate that message into everything: Speak like a cane at our school board meetings. Collect data like a cane when we present. Study like a cane for a test. Talk like a cane with our peers.

Through our district hashtag and consistent sharing of engaging pictures, videos, and other posts, we are hearing positive feedback all over our community. And, we have the Facebook data to back up the increase in engagement, as well.[4]

—**SUPERINTENDENT CRAIG OLSON**, Hayward Community School District, Wisconsin

Creating an Awesome School #Hashtag

Now that you understand how hashtags work, it's time to create one for your school!

If your school is on social media, you *need* a customized school hashtag. Hashtags work on Twitter, Instagram, Google, and even Facebook. Notice that our business name even starts with a hashtag: #SocialSchool4EDU. It's that important!

What is the recipe for an awesome school hashtag? Here are five important ingredients.[5]

1. Make it unique. Choosing a cool hashtag that is already used by other schools will get you nowhere. The easiest way to see if your hashtag idea is being used is to type it into Twitter or Instagram. Is it being used by other schools or profiles? Is it being used in a negative way? If your mascot is a Viking, #GoVikings is used a lot! Using this as your hashtag would not create a flood of posts just about your school.

2. Keep it positive. You want to develop a hashtag that evokes some school spirit. This has been a learning experience for me. At the first school I worked with, we just did #NewAuburn. It isn't negative, but it certainly doesn't evoke a lot of school spirit. Your hashtag could be your mascot, like #CameronComets.

3. Keep it short. Twitter holds you to just 280 characters for each post. Even for Instagram and Facebook, you want to keep updates fairly short. People come to see photos and read quick phrases, not novels! But don't skimp on meaning. When Rice Lake wanted to reference their "Warrior Way," #warriorway was already used by others. To use #RiceLakeWarriorWay would have been pretty long (eighteen characters). They settled with #RLWarriorWay (twelve characters).

4. Use it all the time. People will only start to use it in their own social posts if they know about it. There is no official announcement process; just include it on EVERYTHING. Use it on:

- ✔ Posters in the hallways
- ✔ Clothing and other branded items
- ✔ Athletic and music programs

- ✔ The cover photo on all social media platforms
- ✔ Email signatures for all staff
- ✔ And, of course, use it on all your own social media updates

5. Search and reward! Search for the hashtag at least a few times a week on each social media platform. Grab some of the photos to share on your school page. You can even reward with a simple prize or other recognition like "#ColfaxPride Student of the Week."

Special events can certainly have customized hashtags, as well, but encouraging the use of one hashtag for all things related to your school will be extremely helpful in getting your story out to the world. There isn't a formal reservation process for your hashtag. You just start using it.

I also recommend selecting a hashtag and sticking with it from year to year. You can have annual slogans or goals for your school or district, but your hashtag shouldn't change. When you pick one to use consistently, you'll see the story grow within your community and beyond.

15 HIP #HASHTAGS

Get energized by this list of stellar school hashtags. If your school needs a refresh, let these inspire you![6]

#LifeAtThe803 – This one from Wheaton, MN, sounds like it could be the name of a TV show, right? Students got involved in coming up with this great hashtag.

#BoostTheOost – When you can add a rhyme in the hashtag, it makes it twice as nice! Oostburg School District (WI) came up with this one.

#MosaicOfMinds – Glendale Elementary School District (AZ) provides educational services to more than 11,000 K–8 students in seventeen schools.

#OurHighlineStory – The Highline Schools story is not a single story; it is comprised of multiple stories told by their staff, students, and families.

#Proud2BPirates – Belton School District #124 (MO) has a unique hashtag that's easy to share in a sentence. It's always great when the hashtag evokes feelings of pride!

#EdgarExcellence – Does your mission statement include one standout keyword? Include it in your district hashtag! That's what Edgar, WI, did with their school hashtag.

#ATeam206 – Alexandria, MN, also has a great logo with an easy mission: "Rich Tradition. Bright Future."

#HCPSfamily – HCPS (NC) brings teachers, staff, students, and families together on social media to show what makes it an amazing place to work and learn.

#PointerNation – Many districts use the word "nation" in their hashtag, but Mineral Point, WI, is telling a big story in their district of just 700 students.

#WEareONE10 – Unifying a district made up of five different communities isn't an easy job. Waconia, MN, leveraged a big rebranding initiative a few years back, and it has paid off!

#LikeACane – The Hayward Hurricanes (WI) incorporate their hashtag into everything they do, and they get their community involved, too!

#Celebrate138 – The goal of your social media efforts should be to celebrate your students and staff. This hashtag from North Branch Area Public Schools (MN) makes perfect sense!

#GrowCaroline – Caroline County Public Schools (MD) has a small rural school system with ag in their soul. Their tagline is "Growing the essentials for learning."

#AvonExperience – Avon Community School Corporation (IN) has the tagline "Educate. Excite. Excel." They also want to share the experiences of all their staff and students.

#MiddieRising – Elizabeth Beadle, former Communications Director for Middletown City Schools (OH), describes their hashtag: "We're in this together. We're better together. We're one together. We rise together. We are #MiddieRising."

Grab It!

Need more great ideas? Grab the blog post titled **"10 Cool #Hashtags for Schools"** from the free bonus resources at socialschool4edu.com/book.

YOU'VE GOT THIS!

- ❑ Understand why and how to use hashtags as an important aspect of telling your school's story.
- ❑ Choose one hashtag that represents your school.
- ❑ Avoid popular hashtags that would not point people to your school.
- ❑ Make your hashtag unique, positive, and short.
- ❑ Use it everywhere.
- ❑ Search the hashtag and reshare great content.

GRAPHICS TIPS AND TEMPLATES

8

Presenting a professional image to your community starts with branding and graphics. Let's not settle for just OK. Let's help your district shine!

Case Study

Meet South Washington County Schools: Otherwise known as SoWashCo Schools, South Washington County Schools is a suburban district of almost 20,000 students, encompassing twenty-four schools, just outside of the Minneapolis–St. Paul metro area.

The challenge: SoWashCo Schools leadership determined it was time to refresh their branding and messaging to show cohesion across the district and to reflect the student voice. Leaders believed their existing branding wasn't highlighting their important values. Students said the long-time mission statement of "igniting a passion for lifelong learning" didn't resonate with them anymore.

The process: SoWashCo Schools partnered with a Minneapolis company, CEL Marketing PR Design, for a brand refresh. The CEL team created a new brand logo that is a simple wordmark in bold, bright colors to present the school district as modern, fun, and friendly. The messaging and visuals were developed to represent and work with communications to students, teachers, and staff and to resonate with external communications about the district. CEL selected colors and patterns that represented values that are important to SoWashCo Schools: accessibility, relatability, collaboration, and unity. Each of the high school's colors is represented in the palette for district cohesion and equity. The new tagline is simple and clear: "Be seen. Be heard. Be bold."[1]

CASE STUDY

The outcome: The new branding and messaging showcases SoWashCo Schools as a district that puts students first. It now "reflects and amplifies the vibrant, welcoming nature" of the school district.[2] Thanks to the rebrand, the district has an extensive set of graphics available in its style guide. That includes a library of colors and patterns that fit with the graphic identity system, a "moodboard" with branded examples for photos and posters, and pre-approved artwork. The district communicators are set to create outstanding cover images and graphics!

Grab the link to the SoWashCo Schools brand guide for inspiration in the bonus resources for this book at www.socialschool4edu.com/book.

so wash co SCHOOLS

Profile pictures, cover photos, and graphic images – oh my! When it comes to making decisions about your public image, branding your social media channels matters. I can tell with just a glance at your Facebook or Twitter page whether you are an amateur or a pro when it comes to branding. But don't lose hope! It can be fixed.

Now is the time to take action to ensure you're following the basics for good branding. The following practices are a top priority when it comes to your images.

Facebook and Twitter Cover Photos

If your cover image is a photo from five years ago, it's time for an update! Your Facebook and Twitter cover images are prime real estate when it comes to showcasing your school and grabbing attention. I have seen many schools with no Twitter cover image at all. That's a big error! A few quick guidelines on cover images:

- Make sure your photos are high resolution.
- Do NOT use a stock photo. You need photos of *your* students and staff.
- Consider branding a collage with your colors, photos, and other information.
- Make sure you create your cover images with the right dimensions. You will want to check the cover image on both desktop and mobile. At the time of writing this, the top and bottom of the Facebook cover image will be cut off on a desktop. You also want to pay attention to how the profile image covers up part of the cover image when your followers view your page. If you use any words, make sure they appear on both mobile and desktop.

If you have a call to action with your cover image, remember that you can include a caption on the image.

- As you create cover updates for each profile (Facebook, Twitter, and YouTube), you'll need to alter the dimensions. I urge you to stay consistent across platforms, so as you update your Facebook cover photo, update the image on other channels, too. It just makes your job a bit easier so you're not trying to track different cover image ideas for each.

COVER PHOTO TIPS

You can use your Facebook, Twitter, and YouTube account cover photos to brand your school and tell your story. Here are my tips when it comes to your cover photos:

- ✔ Change out your cover photo at least once per quarter. You should match it with the seasons. So don't include photos of students in winter gear during the summer months!
- ✔ Represent a variety of ages, genders, and activities on your cover photo.
- ✔ If you use just one image on the cover photo, then you should think about changing it out a few times a month.
- ✔ Include your school hashtag on the cover photo.
- ✔ If you have a special award you'd like to promote, consider adding it to your cover image.
- ✔ Make your cover photo include a call to action! To add this on Facebook, you simply click on the photo once you've uploaded it as your cover photo, click "edit," and then add a caption to the photo.[3]

Your Profile Image

Your school's profile image is important! It should be an image with very minimal text. Your logo in the profile image should be clean, consistent, and look great in the square or circle. You may have several versions of your logo, but this version should be the most simplified form. And I'm not talking about a pixelated logo that doesn't really fit in the circle. A tagline or the spelled-out name of your school in small print is *impossible* to read on your social media channels.

Remember: the name of your social media account will always be next to the profile image when you post, so your profile image doesn't need to include the school name! For some schools, your logo will be your mascot.

For others, it will be your district logo. Your profile image should not be a picture of your school or a photo of students or staff. You stand out from other profiles by using your logo. If your logo poses a challenge for this standard, it may be time to create a new logo.

Take a look right now at your pages. All social media channels should have the same version of your logo.

Your logo in the profile image should be clean, consistent, and look great in the square or circle.

The Profile Picture Debate

I just made the case for having your logo as the profile picture. And I've often boldly declared that this image should never change. However, one school district provided a convincing example for an exception.

Peel District School Board (@peelschools) has created seasonal versions of its logo that are relevant to the time of year but still true to the primary logo. While the design and colors changed slightly, the main elements are still recognizable.[4]

So how can this help your social media presence on Facebook (and beyond)?

Show up in the newsfeed. With each new photo you add, you have the chance to show up in your fans' newsfeeds. By changing up your logo to something seasonal, you are likely to draw in some "likes" on the image, and that will, in turn, cause Facebook to deliver it to even more people.

Add personality to your district. Schools are run like a business. They play a critical role in our society. But showcasing a fun side is important on social media platforms like Facebook, Twitter, and Instagram. Participating in trends, events, and holidays provides another chance to get in on the conversation with your community.

Provide shareable content. With Peel District, many fans steal the photo for their own profile pictures! This is the exact purpose of social media. Turn your fans into the influencers in your community that help spread the word and message of your incredible school.

If implemented well, schools can effectively change their profile pictures to attract even more positive attention to their mission of celebrating their students.

Grab It!

The best way to learn is with specific examples, right? We have a blog full of helpful tips and tricks to get your graphics on the right track. It includes images to show you exactly what to be aware of. Look for "**Social Media Graphics: Just OK Is Not OK**" in the bonus resources at socialschool4edu.com/book.

Do a Graphics Checkup

What makes a "good" graphic? Is it OK to be just OK? Allison Martinson, graphic designer at #SocialSchool4EDU, shared five problems to watch out for, plus solutions for each one.[5]

Problem 1: Using Google Images clipart

Isn't it amazing how you can type anything into a Google search and, BOOM, it's at your fingertips? Sounds dreamy, right? Here's the problem: the results are usually copyrighted, low quality, or too small to use effectively on social media.

Solution: Make your own graphics! Apps like Canva, Pic Collage, and Word Swag allow you to utilize photos, add custom color backgrounds, and add your logo to make the perfect graphic. Just download your creation to your smartphone or computer and easily attach it to your social media post.

Problem 2: Overlaying text on a busy photo

Adding text over the top of a photo can be tricky if the photo is busy or has high contrast. If the text color doesn't pop, it's likely that it'll be hard to read on a screen.

Solution: Place text over a "dead area" in the photo using a contrasting color or place a solid or semi-transparent box behind the text in a contrasting color over the photo. Another way to remedy this is to use a filter on your photo that minimizes contrast and make sure your text overlay is done in a contrasting color.

Problem 3: Using a photo of your building as a cover photo

"Nothing stirs emotion in me like an aerial view of a giant brick building," said no one ever! Buildings can't smile, laugh, or convey a sense of community. Therefore, they're not what you want to use as the face of your social media pages.

Solution: Always use photos of students and staff on your cover photo. Keep inclusiveness in mind when selecting your photos. Use photos that show your school's range in age, race, and status. Photos that make you smile are always the best choice!

Problem 4: Using too much information on your graphic

Do you ever see a poster with SO MUCH INFORMATION on it that you have to look away? This same thing applies to social media. People scroll through their feeds quickly, and if your graphic is chock-full of information, it will make their eyes roll back in their heads and possibly cause a blackout. That's an exaggeration, but they probably won't click on it for more information.

Solution: Keep it simple. When making a graphic for an event, only add a few eye-catching words and leave the details for the description. See? Keep it simple!

Problem 5: Inconsistent branding

This problem is probably the most common of all. Many schools have endless logo variations that have popped up through the years, and no one knows who made them or which one is "the right one." And do you know how many shades of blue there are? Just saying "blue and yellow" doesn't cut it.

Solution: Get your branding on track! We've covered this already. Using consistent fonts, colors, logos, and graphics creates recognition, and recognition creates a sense of community for your followers. Get to know your district's hex color codes and use them consistently on graphics. Consistent branding will, well, *brand* your district into your follower's minds.

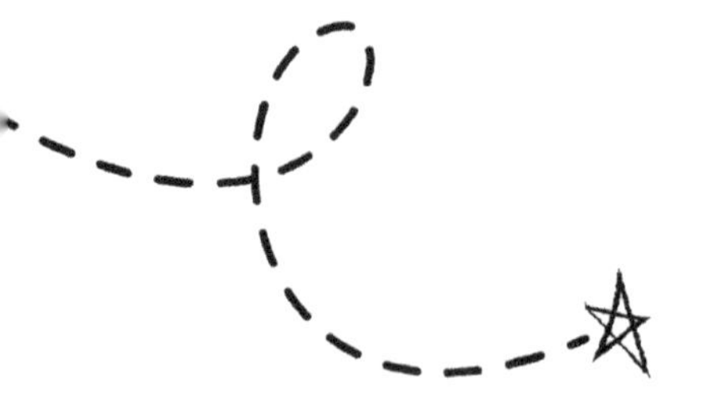

TIP: OFFER TO HELP

MACI STOVER, communications director for Clinton Public Schools in Oklahoma, offers to create graphics for staff to keep them on brand and makes it easy for them to submit a request. At the start of the year, she puts her business card with two QR codes on the back in the staff members' mailboxes. "One QR code takes staff to our 'good news' form, and the other QR code is for staff to submit a graphic request. If they have an event coming up that they want to advertise, I can make a graphic for them," she said.

Maci strives to stay true to her style guide when she's designing. She does what she can to make designs consistent and to avoid having "too many variations" of their logo. Maci's design goal is to make sure anyone scrolling social media immediately recognizes the school's branding, just from the color and style.[6]

Making a Graphics Kit

Every post on social media needs an image. No excuses. But there's a solution! Create your own graphics kit with a few templates that you can use, and it will make your life easier as a social media manager for your school. Here are eight common graphics you can create and have at your fingertips for easy posting.

1. Thank you
2. No School
3. Mark Your Calendar
4. Important Info
5. FYI
6. Weather Update
7. Join Our Team
8. Cancellation

If you have special school-specific occurrences that happen repeatedly, you can make adjustments for your school. For example, my school has a two-hour late start that happens each month. I made a custom graphic to use each time I share that reminder.

In addition to the eight common graphics above, you should leave one completely blank. Using a font similar to your standard graphics, you can customize your message when other types of information needs to be shared. I've done this for parent-teacher conferences, volunteer requests, and livestream details.

Once you have your kit made, you can save the images to a cloud storage folder – like Google Drive. That way, you can grab them from a phone or a computer when you need one.

I mention Canva often in this book because it's easy to use and has tons of free templates you can customize with your school colors and branding. Plus, it offers the ability to quickly resize an image for different platforms. In addition to the graphics listed above, you can create custom templates to use as photo frames for your content.

READY-MADE GRAPHICS

NATALIE EITING is a special education teacher for the School District of South Milwaukee who also serves as one of the social media managers in the district. Natalie said that having the ready-made graphics has helped them "raise the level of the quality of what we're putting out efficiently, and it keeps a similar tone." The posts look professional, and it has helped with consistency on the posters people create, as well. Natalie said, "We're just teachers who pretend like we're graphic designers, but we made it work."[7]

Grab It!

Need an example of a set of graphics? We have them in a short YouTube video linked in the free bonus resources at socialschool4edu.com/book. Look for **"Simplify Social Media with These 8 Reusable Graphics."**

PRO GRAPHIC DESIGNER AND CANVA NINJA

Elishia Seals is the former public information director for USD 250 Pittsburg Community Schools in Kansas. Elishia is a trained graphic designer, but she knows that many school communicators are not. Professional design programs such as Adobe are the elite software for design, but they require a learning curve that most school communicators don't have time to tackle. For professional designers, it might also be difficult to get past the resistance to using simpler software. Yet many users don't use the tools to the fullest.

Elishia said, "Why wouldn't I make my life easier?" She discovered that Canva was faster than her professional design programs, so now she uses it every day. She set up folders in Canva, and she works to keep the content organized by designating time weekly to sort through finished graphics.

One of the biggest mistakes even Canva users make, Elishia said, is "they're not customizing their templates." Templates are a starting point, and the user then adds their brand touches to it. She also explained that in Canva, users can build a branding guide that others on the team can use. The branding guide saves things such as colors, logos, and fonts.

A bonus Elisha discovered when she helped others in her district with design is that she could become a support resource. When someone handed her a ten-year-old flyer and needed help with updating it, she worked her magic by scanning a PDF into Canva and letting it convert the file to something editable. She said, "I love when someone hands me something and says, 'I started this, but I know that you can add elements to this that will make it better.'" That's her favorite thing to do.[8]

Be sure to include your hashtag on your graphics!

Story Templates

Another great place to use templates is for Instagram Stories. The dimensions should be 1080 pixels wide by 1920 pixels high for Stories. Canva has tons of free templates you can customize with your school colors and branding. I recommend keeping it simple with just your school colors and maybe a logo. You want this to be a nice background to any text you want to add while in Stories.

Save your graphic to your phone and organize it into a folder so you can easily find it every time you need it. Then, add the photo to your story, and you're all set!

These work great for general announcements when you only want to add some quick text or GIFs to communicate a message to your community.

If you want your announcements to really pop in Stories, take the time to design your graphics to include all your text and images. This also allows you to use any fonts, photos, and colors you choose. This is a great option if you have something really important to communicate.

Caution on Overbranding

Now that we've covered the importance of looking great on social media, I need to provide you with a word of caution. In my opinion, there is a case for too much branding on your social media pages.

Not *every* photo or Instagram Story needs to be on a branded background. You should share photos and videos that are not edited in any way.

You only have twenty-four hours in your day (although I think school social media managers deserve twenty-five). You don't want to spend all your time customizing graphics and adding your logo to every single image you share.

My goal is always to make your life simpler – not to complicate it. I say it often: manage social media, and don't let it manage you. Your graphics kit will keep you on brand and efficient. Your school will look great online, and you'll have a life. That's a win!

YOU'VE GOT THIS!

- [ ] Use one consistent logo.
- [ ] Check your profile picture. Is it clean and consistent, and does it fit in the circle or square?
- [ ] Update your cover photos at least once a quarter.
- [ ] Feature people, not buildings, on cover images.
- [ ] Don't settle for just OK on your graphics. Keep them simple, easy to read, and consistent with your brand.
- [ ] Make a set of graphics for the posts you make the most often. Put those up whenever you need to make an announcement about weather, athletics, job postings, and more.
- [ ] Trust the simple tools such as Canva to get the job done without a lot of special training. Make your own branded templates to use.

STORYTELLING

STORYTELLING

GETTING AND ORGANIZING CONTENT

9

Social media plays an even bigger role in your district than ever before. But you can't do it alone. Creating an army of storytellers will help you celebrate all the awesome things happening in your school and give you confidence that you'll always have great content.

Case Study

Meet Steffanie Stratton: Steffanie is the communications representative for Northeastern Local School District (NELSD) in Springfield, Ohio. She is the sole person in the communications office in the district of 3000 students in K–12.

The challenge: When Steffanie gets content sent to her own email address, it becomes lost among tons of other messages. And she has to mentally process: Where is it? Did I forward it? Did it get buried? "I struggle with being all over the place because we have so many hats," she said.

The process: Steffanie set up a separate email for receiving content: socialmedia@[schooldomain]. She said having it all come to that inbox rather than her personal one provides her with a time-stamped list of submissions that she can filter into folders and categories. She checks the email once a day and organizes it into folders for future posts. What happens if staff forgets and uses Steffanie's email instead of the social media one? She continues to kindly reiterate the process.

The outcome: In addition to the benefit of keeping her email separate from the social media content, Steffanie discovered that having a separate social email was helpful during a transition time. She and a former team member were able to both have access and collaborate during that transition, and Steffanie still had access after it was completed.

CASE STUDY

All the time-stamped emails and photos were still there for her to access. She said another benefit of the socialmedia@ address was staff not having to remember how to spell her name (which has an unusual spelling) when composing a new message.[1]

celebrate YOUR SCHOOL

ONE STORY AT A TIME!

When it comes to getting social media content, you can't do it alone. You'll need to get your staff involved in the process of gathering content for you. Using this method, everyone can feel part of telling your great story. Make sure all your teachers, support staff, administration – everybody – knows that they can tell the story.

Make sure all your teachers, support staff, administration – everybody – knows that they can tell the story.

The second thing you need is a simple way for them to get content to you. Have them send you one or two things per month that you can post on social media. How do they get that to you? Set up an email address like socialmedia@YourSchool.org and direct everything through that email.

Why an email address is the best way:

- ✔ It's easy.
- ✔ It's one process for the person sending (no forms or drop boxes).
- ✔ It lets you stay organized (nothing comes to your texts or personal inbox – or multiple locations).
- ✔ If you go on vacation, someone else can check that email address.

You might be asking yourself, "Andrea, but what if someone sends in something that I just can't post? The photo is really bad, or it speaks of something that might cause controversy in the comments?"

You definitely need to be ready to say no to some content. Proper training will help with this, but I have dozens of stories where we had to tell the staff member no.

One time, I received cat dissection photos. I can do frogs, cow eyeballs, and even pigs. But I just couldn't share photos of cats being dissected. I thanked the staff member but explained that sharing this could lead to some big tears from cat lovers and big shouts from PETA.

I rarely have someone tell me that they have too much content, but if that is the case for you, you can always post some content to Stories. This won't clog up your page feed, and you can keep the really awesome content for your feed.

Checklist for Getting Staff Involved

Ideally, you should devote at least fifteen minutes during staff back-to-school sessions to talking about the importance of social media. If you missed that window, ask for time at the next staff meeting.[2]

Your title may include a reference to communications or public relations, but you are not the only person who can take photos to be used on social media. Are your photos going to be better than what others take? Probably. But you can't be in every location within your district to capture the stories that deserve to be told. So, if you get pushback from staff not wanting to send in content because "that's not my job, that's your job," you can use this argument. You also need to keep in mind that you won't get everyone on board. That's OK. Train your staff anyway.

This is my quick checklist of things that my team and I cover with staff at each of the schools we serve. I suggest using a slide deck to accompany your discussion.

- ✔ Social media is a powerful way to reach your community. Start by sharing a few statistics that show the reach you get on your district's Facebook, Twitter, and/or Instagram pages. You can also include a few positive comments that have popped up on posts that speak well of the staff or school as a whole.

✔ Share your social media channel handles and district hashtag with your team and encourage them to follow and interact with the content shared. The community sees that interaction!

✔ Remind staff of the method they should use to submit content for social media. This includes photos, videos, and brief descriptions. They'll use the social media email. Ask staff to ONLY submit content through this method. If you receive content through text messages, Facebook messages, and emails to your personal email address, it's more likely that you will miss something, and you're definitely going to feel burned out.

I have met many social media managers over the years who have ignored this tip. They are just happy to get content, so they'll take it any way they can get it. Trust me – providing one simple method and correcting staff if they stray from it will save your sanity! Pick a method and enforce it. Some schools do use a form, and they have found a way for it to work, but I still feel a separate email is easiest.

✔ While talking with them, ask them to send a test email – right now – to that email address. The advantage of doing this is that the next time they need to send in an email, they just start typing the first few letters of your email address, and voilà, the social media email address auto-populates! You may also want to print out a prompt that they can place at their desk. The prompt will contain the email address and remind your staff to include details. For example: Today we are learning __________ by doing __________.

✔ Challenge staff to provide one to two social media submissions every month. Remind them that you need to celebrate ALL of your students and that if everyone participates, you can make sure that no organizations, age groups, or activities are left out. We'll cover some ideas in the next chapter.

- ✔ Not all photos are created equal. Ask staff members to email you the best photo quality possible and a short description. If they have 400 photos from an event, do NOT send all 400 photos to you. Pick the best handful of photos that highlight the event. You could even show them examples of good photos vs. bad photos! We'll talk about tips for better photos later in the book.

- ✔ Know who may not be photographed. At the start of the year, there is likely a new opt-out form that families will complete. Remind staff that the office will be redistributing that list after registration and that they should avoid sending in photos of those students.

If a student appears in a photo that is not allowed to be used on social media, do not place an emoji or sticker over their face. This draws more attention to that student. Crop that student out of the photo or use a different photo.

- ✔ If staff members have their own Facebook page or Twitter account, awesome! But you can't catch every great post that they share (nor do you want to – you don't need to know all the small details from their accounts). Ask your staff to email you the URL link to important posts that should be celebrated on the district account. Also, make sure they use the district hashtag in their posts!

- ✔ Assure staff that you will reply to email submissions they provide so they know you received the content.

- ✔ Remind staff that not all content is posted the day you receive it. The algorithm is a real thing, and you have a strategy behind your approach to posting photos, videos, and stories. It will be posted in a timely manner – but you might have to wait![3]

DOES YOUR STAFF FEEL AWKWARD ABOUT SNAPPING PHOTOS IN CLASS?

If your staff is worried that having their phone out in class to take a picture will be intrusive, here's a tip:

> Teach them to snap the pictures with their phone but wait to send them or post them until later. They don't need to be *posted* in the moment!

WHAT IF STAFF SUBMIT PHOTOS THAT AREN'T PRO-QUALITY?

In a podcast episode of *Mastering Social Media for Schools*, Sara O'Donnell said she has changed her stance in the last few years on what was once important for her. She now advocates for "letting go of the quality that you may get [for photos]. . . . I want the highest quality, but it's OK that I'm not the one taking the photo with my [professional camera]." Rather than seeing it as lowering her standards, Sarah said she wants to "encourage and appreciate and then celebrate" when she receives content after she requests a photo of something happening in the classroom.[4]

—**SARAH O'DONNELL**, M.S. IMC, Director of Communications, Stevens Point Area Public School District, Wisconsin

- ✔ We don't see a lot of interaction with our posts on weekends, so we normally only post during the week. Let staff members know if this is your plan. If you don't follow this practice now, you might want to start! What a relief it is to know that you don't have to monitor comments on a new post that goes up on a Saturday or Sunday! And please know that if there is a big game or contest on the weekend and you want to share results right away, you can.

- ✔ Set good boundaries with the times you monitor the email address. We stress the fact that the email is monitored during school hours. If something is sent after school hours or on the weekend, we will respond the following school day and schedule the post from there.

- ✔ Since we are contracted with schools to manage social media, we have internal contacts at the school that can post in case of emergency (no school days, early releases that are unexpected, etc.). We remind them of who that "urgent post" person is. You may want to remind your staff about others that have access to your social media pages. There should always be a backup if you are not available or on vacation.

- ✔ We also remind staff of Facebook Live, Instagram Live, and Story opportunities. Getting help from a few staff contributors will make your job much easier. One key person in each building could help, but you'll need to provide some training. The resources in this book and my membership program can help with that! [5]

Grab It!

Need a little extra help with Facebook Live? We have two resources you can download from the free bonus resources at socialschool4edu.com/book:

- Make it happen with an easy **Guide to Facebook Live: A Step-by-Step Checklist.**
- Spark your imagination with **30 Ideas for Facebook Live**.

Need help with photos? You may be posting on social media on behalf of your school, but that doesn't mean you need to take every photo! Help your staff help you. Grab the resource called "**10 Tips for Better Photos for Social Media**."

A Social Media Directory

Now that your staff understands the process for *your* posts, how will you track the content on *other* channels? In chapter 4, we discussed staff-run pages. These include team pages, student council pages, and more. If you have a large district, you may also have school buildings running their own accounts. Does keeping track of the various social media accounts that are connected to your school sometimes feel as if you are herding cats?[6]

I get it.

One of the biggest concerns of school communicators is the vast number of social media accounts that are out there – some that they don't even know exist! Let's walk through a process to wrangle those cats and create a social media directory that can live on your website. This requires creating a process, not simply a one-time project, but you'll have many content sources once you set up this directory.

Step 1: Create guidelines. We covered those in chapter 2. This will create a framework for what is and is not allowed when it comes to these accounts.

Step 2: Create a submission form. The next step is to get organized and create a submission form for staff, advisors, and coaches to complete in order to have their social media channels listed in your website directory.

You can advertise it this way to encourage people to complete it: "Spread the word about your social media page for your club, team, or classroom! Submit this short form to get listed on our social media directory."

Your goal is to help people understand the responsibility that comes along with managing a social media account on behalf of the school. This submission form should also be included in new staff orientations.

INFORMATION SHARING AND AUTHORIZATION FORM

This example comes from Eudora Schools in Kansas. Staff is asked to submit the following information:

> Those employees wishing to create and maintain a Facebook or other social media profile as a part of their job should complete the information sharing form below. (This includes pages for classes, teams, organizations, or groups of parents and/or students. It does NOT include personal profile pages.)
>
> Employees setting up and/or maintaining a Facebook page or group also agree to add a supervisor and/or [Communication Person] as an administrator for the page. In case of an emergency that prevents you from being able to access your page, this will ensure timely and accurate information is shared with your followers. Adding a page administrator is simple [links to instructions].
>
> Completing this form will connect you with social media champions in your building and will allow the district to maintain a central directory of social media profiles in our schools to help parents and patrons.
>
> - My name:
> - My email address:
> - I plan to represent the following class/team/club/project/etc.:
> - On social media the social media tool I intend to use is:
>   - Facebook

  - » Twitter
  - » Instagram
  - » Other
- The URL for my page (or my Twitter handle and/or hashtag) where some can find it is:
- Other MAIN audience not listed above:
- Here is a description of the type of information or content I plan to share on my site:
- I plan to use the following privacy structure for my page.
  - » Open public access to view and post
  - » Open public access to view, with no ability to post
  - » Viewing access is limited only to invited members (private or secret Facebook group)
  - » Other
- I affirm that I have read and understood the social media guidelines for employees
  - » Yes
  - » No
- I acknowledge that I will be solely responsible for managing the information and posts on my page. I also affirm that I intend to give one of my supervisors, and/or [Communication Person], administrator-level access to my page, group or account, to assist in case of emergency.[7]
  - » Yes
  - » No

Step 3: Review all submissions. As submissions start to roll in, you will need to set aside time to review the responses. If you see any red flags, you'll have to get back to the submitter. If it is approved, you can proceed to the next step.

I don't recommend having the submissions reviewed by a large group of people. If you are responsible for communications, you should be empowered to make these decisions. You may want to run the names by the leadership team in a meeting to make sure they don't have any concerns, but this step does not require a large committee.

Step 4: Create a directory on your website. Now that you have approved pages that are associated with your school, let's give others an easy way to find them! The best spot to publish a directory is on your website. This allows district stakeholders to find the official social media accounts associated with your school.

You might group the list by departments or schools, use icons or text links, or use charts. You can get as fancy as you want or keep it very basic. It's totally up to you! It just needs to be functional – with real links to the social channels referenced.

Remember, this is going to be a living page. You'll be updating it periodically, so don't worry about having it be 100 percent complete before you "go live." Just get that page published, and you can always add to it later.

Step 5: Share the directory with your staff and community. Once you have the directory ready, it's time to promote it! You can announce the newly created directory via social media, of course, but you can also share it in your newsletter, talk about it at your next in-person event, and announce it in staff meetings. If you have a local newspaper, you might think about sharing it there. The next time you send out a direct mail piece, add it to the communication.

The point is that you need to let people know that you created this amazing tool to help your community stay connected!

Step 6: Set up a quarterly review. Regular reviews of the directory are important. You can choose to do this once per quarter – or at a minimum, every six months. I wouldn't leave it to once a year because there is just too much that can happen in that time frame.

This periodic audit helps with a few things.

First, you need to verify that all the links still work on your website. Staff members leave, and pages are shut down, so you want to make sure that all listed links are working.

As you test each link, you should also review the content being shared. Is it meeting your social media guidelines? Do you spot any red flags? Things that might cause you to follow up with the account manager include:

1. Lack of posting
2. Poor spelling/grammar
3. Inappropriate comments without response

Another important element of this review is to remind yourself to reach out to all staff to gather new social media channels. Simply share the form that they must fill out to have their social media channel listed in the directory.

Step 7: Offer continued professional development. Social media is always changing. If your school or district is allowing your staff to use it, then you should also invest in the time to offer professional development. Once a year, offer training to teach best practices.

This training could be held either in-person or virtually. If you need help, the team at #SocialSchool4EDU facilitates small-group training for internal teams! We have customized sessions that we can do for your group.

When it comes to preparing your training agenda, keep these items in mind:

1. Spend time on common "dos" and "don'ts" for social media.
2. You could show some examples of your school's existing channels and identify good examples of proper social media use. You can also find examples that might need some help – although you don't want to call anyone out in front of others. Provide that feedback one-on-one.
3. Your door should always be open to answering questions about social media, but this training session should also leave plenty of time for questions and troubleshooting.

It won't happen overnight, but once you create a directory and stick with the process, you will have more peace of mind and more positive, online celebrations for your school!

HOW ERICA BOOSTED STAFF INVOLVEMENT

ERICA LOIACONO is the director of public relations at Community Unit School District 200 (CUSD 200) near Chicago. Her district needed a boost in community engagement. The board challenged its school leaders to find ways to demonstrate to the community that their investment in CUSD 200 was worthwhile.

So, Erica launched a social media campaign called "Telling Our Story, One Classroom at a Time." The goal was to get as many staff members on board with the campaign as possible. Erica sought to share more positive stories about what was happening at CUSD 200.

But instead of sending an email that would likely get buried in their inboxes, Erica hosted sessions to highlight the impact social media could make in achieving district goals, plus how individual staff members could get involved with these efforts.

These were optional sessions, but guess what she did to encourage good attendance?

She provided lunch, of course! Never underestimate the power of free food.

In Erica's case, the district's new food service provider donated sandwiches and chips, and her department provided water and cookies.

Erica started off with a brag session, highlighting district rankings, test scores, and positive parent reviews. Then she explained how social media could play a big role in accomplishing the board's initiative of increasing community engagement. She made her case for social media and then shared real-life stories, showing screenshots of great social media posts from the district, personal testimonies, and more.

"For a lot of our families, it might be difficult, depending on their circumstance, to stay connected and stay engaged at school. Social media crosses all socioeconomic barriers. But you guys are where it starts. Parents just want to see a little bit more of what that education actually looks like so that they can connect with their child!" she said.

Then she invited staff members to help tell the district's stories by liking, commenting, and sharing school posts. She encouraged them to submit content or even run their own social media account.

At this point in the presentation, Erica had her listeners bought in and ready to help.[8]

YOU'VE GOT THIS!

- [ ] Create an email – one place – where staff can send content.
- [ ] Ask for one or two posts per month from each person.
- [ ] Encourage staff to follow social media profiles and like, comment, and share.
- [ ] Do a social media orientation for staff at the beginning of every school year.
- [ ] Train staff on the types of content that work best.
- [ ] Create an organized system for tracking staff pages and include a directory on your website.
- [ ] Ask staff to send you links to great posts on their pages.
- [ ] Review the social media directory quarterly to keep it up to date.

CONTENT IDEAS 10

What if you had so much great content that you didn't know what to do with it all? What if 5,000 or more people saw your posts every week? This isn't out of your reach!

Case Study

Meet Callen Moore: Callen is the public relations officer for Walton County School District in Georgia, a district with fifteen schools and more than 14,600 students. When Callen started out in the district, there were 3000 followers on Facebook. At the end of 2022, it had nearly 12,000. She also added Twitter and Instagram to the school's channels.

The challenge: Figuring out what resonates with followers on different platforms and engaging without reinventing new content for each channel.

The process: Callen has learned to focus on where the best audiences are. She still cross-posts to all platforms to be more efficient. However, she's observed that content doesn't have the same effect on all platforms. She added a "question of the week" feature and has observed how it's received on each platform.

The outcome: Callen has had amazing results on Facebook with the question of the week. For example, within five hours of asking, "Who was your favorite teacher?" one post had 296 comments on Facebook. People tagged teachers, and teachers engaged back. Callen posted the same question on Instagram and had no comments and only a few likes. She said she knows Instagram is more picture-based, and the text-based content may "not have picked up as well." Posting the content in Stories and prompting with a question can

CASE STUDY

potentially bring engagement that a regular post doesn't have on Instagram. Either way, she serves up the same content but continues to find ways to present it in ways where it can have the best effect.[1]

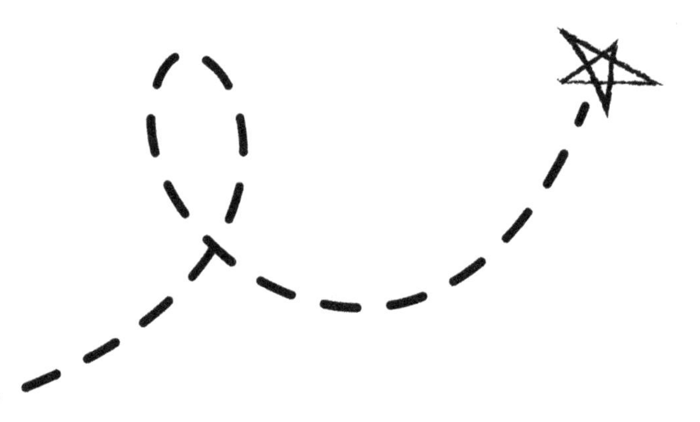

Do you know the number one concern from nearly every school I talk to? Hands down, this comes in first place by a mile! Content. How do we get content to post on social media?

I'm a bit amazed that this is the number one issue. When I walk into a school, I look around and can instantly see ten things that could be shared with the community.

- ✔ A simple photo of a teacher tying a student's shoe
- ✔ A short video interview with the child holding the door for other students
- ✔ A picture of the artwork that a teacher has displayed in the hallway

However, when you're so close to it, it's hard to see. You have a lot of other priorities on your plate, and I understand that it can be tough.

Great social media doesn't rely on one person. Creating awesome content requires a system. It also requires a village of people willing to spend a few seconds to capture the amazing things happening in your school.

And those everyday things are the foundation of your social media strategy. We'll cover feature ideas and other ways to connect with your followers, but it all comes back to showcasing what your students and staff are doing in your school.

It all comes back to showcasing what your students and staff are doing in your school.

Even though we aren't in the schools we serve here at #SocialSchool4EDU, we post on channels such as Facebook more than twice per day. We help the schools' posts be seen by 5,000 or more people each week.

I know you want those same results at your school. Let's look at nine ways to make it happen.

Nine Strategies for Generating Awesome Content

1. It takes a village. Every staff member submits one thing per month.[2]
You can't be in every classroom. You also can't attend every special event within your district. But guess what? Nearly everyone on your staff has the equipment they need to help you out. A phone!

They snap a picture, send it to a unique email set up for social media content, and then you can take it from there.

2. Student contributors. Leverage their talents and perspective.
The most important stories happening in your school should come from those you serve – your students! Make sure you share what they are up to and involve them in the process. Videos created for classroom assignments are great for social media.

Ask your students what they love most about your school, and then share their responses. And after high school, what are their plans? Those stories need to be shared with your community!

3. Crowdsourcing. Follow your unique hashtag and other pages, then pull that content for your pages.
You have hundreds or even thousands of people talking about your school on any given day on social media. Pull those stories together with a hashtag that is unique to your school. You can then retweet those stories or even grab them to share on other social channels.

And if you have classrooms, clubs, or sports teams with their own Facebook pages, follow those in your page's feed! When you see something that is worth sharing with everyone in the district, download the photos and post them on your page (don't simply share the post because the reach will not be as good as an original post).

4. Winner-winner. Create contests with simple awards.
Everyone loves a contest, right? Well, I do! I'm one competitive woman. A simple contest can make submitting social media content fun. We have a

great idea of a staff social media BINGO challenge in the resources. Or you can do a drawing only for staff who submitted content. Prizes can be jeans on Friday or a donated gift card from a local business. A traveling trophy might work, as well. We've seen a fun light-up "hashtag" symbol used for the trophy.

And contests don't need to just be for staff. You can have them for your followers, as well!

5. Keep it simple. Repurpose great content.

You only have so much time during your day. Sharing exclusive content on each social media channel is not required. Instagram is all about visuals, so use the best photos that you share on Facebook on that platform. You may want to utilize the best photos for special promotional posts – or future cover photo collages.

6. Instagram and FB Stories. Empower staff and students to take over.

Around 1.45 billion people are using stories every day.[3] This means that our schools need to be there, too! But it doesn't have to be YOU. Ask a staff member to help share stories. It doesn't need to be daily. And remember, stories go away after twenty-four hours, but you can save them to your highlights! We'll cover staff and student takeovers in chapter 13.

7. Build relationships. Mass emails don't work – ask individually.

If you are low on content, it may be tempting to just put out an email to all staff saying, "I need more photos for social media. Please send them in." But this approach doesn't work. We find that building one-to-one relationships is what really moves the needle.

Be as specific as possible. Talk to the art teacher and compliment him on the amazing artwork on display in the hallway. Ask him to send you a simple description of the unit along with a couple of photos. He's so busy he doesn't take time to even think about it – but once you ask, he will! Developing key relationships with a handful of people in each school building will have a ripple effect!

8. Make it easy. Create a system and remind staff routinely.

Your teachers, administrators, and support staff are busy people! You need to make it really easy to submit content for social media. (Remember the suggestion I shared in chapter 9 about setting up one email address for content.)

If you give staff too many options (email, text, complete a form, or send a Facebook message to you), you'll create confusion. Pick one method, and then remind them frequently in staff meetings, staff newsletters, and other ways you communicate.

9. Show them the impact. Measure your efforts and share it back. You must share your social media wins with your staff and school board. If you have a post that reaches 10,000 people, let them know about it! If you receive great comments, make sure to highlight them. Don't assume your staff reads every post on social media.

That report card I'll mention in chapter 17 is so helpful. If they see the impact you are having, they will likely participate more in the future.

....

There you have it – nine proven strategies for more content for your district's social media platforms. Which strategy are you going to utilize first?

BOLD GRADS TO BOLD TEACHERS

NICOLE VALLES, digital content specialist for Barrow County School System, Georgia, was searching through an old yearbook for some #ThrowbackThursday posts when she made the connection that many of the students in that yearbook were now teachers at the school. "I emailed each one of them and asked them if they could send me a picture of their senior year and then a current picture and a little blurb about their history," she said. For fun, Nicole included some of the superlatives from their senior yearbook (one had been voted most dependable and now has been teaching in the district for over twenty years).

Nicole named the campaign "Bold Grads to Bold Teachers." She posted the images weekly, but she did it in a way that kept them all organized in an album. Each time she added a new photo, she put it in the album so the newest photo showed as a post, but followers could scroll through the album to see previous photos.[4]

Grab It!

Want an easy printable to remember the nine strategies for awesome content? Look for the resource titled **"Social Media Made Easy: 9 Strategies for Generating Awesome Content!"** in the bonus resources.

Getting More Photos for Social Media

Sometimes, as school communicators, we forget that not everyone knows how to send in their stories, Heidi Feller said in a guest post for #SocialSchool4EDU. Providing regular tips and tricks to your staff members and even offering to spend a little time helping someone can make the difference between a few people sending you things and everyone sending you a story.[5]

Heidi said, "I once showed a teacher how to install Google Drive on her phone so she could send pics for social media posts to me . . . in the produce section at my grocery store."

Heidi told this not to illustrate her tech knowledge but to bring us in on a little secret – your staff may not be sending you pics and stories because **they don't know how**!

Most people know how to take photos with their cellphones. That's pretty easy. But what about emailing those photos to you? Let's be sure they know the next steps!

TIP: CHOOSING THE ORDER OF PHOTOS

When posting multiple photos on a post about an event, always post the best one first. This is the one that is the clearest, is the most interesting, has bright colors, and has someone's face on it.

We want teachers, bus drivers, food service workers, and paraprofessionals to send us photos they snap with their phones so we can tell our district's story. Celebration is the name of the game, Heidi said.

Here are some ideas you can share with your team to help them use technology to its fullest potential.

- ✔ Create a short video with a step-by-step tutorial on how to send pictures directly from a phone.
- ✔ Add a slide to your next staff meeting that shows screenshots of how to send a photo.
- ✔ Have everyone at an in-service meeting send a test email.
- ✔ If you notice someone struggling to send videos or snapshots, email them and ask if you can pop into their office or classroom for a hands-on lesson. Not only will you be teaching someone, but you will also be enlisting that person as part of your storytelling army by going the extra mile![6]

100 Ideas for Posts

Now that you have the creativity going, could you use even more inspiration? Here are 100 ideas that you could implement for telling your story with specific inspiration for teachers, administrators, coaches, features, and events. These suggestions can be done as photos, videos, collages, slideshows, and more!

Teachers

- Video on why they love teaching
- Sharing a unique talent (such as juggling)
- Selfies with students
- Weekly teacher feature
- Professional development stories
- Celebrate employment milestones
- Honor retirees
- Tips for parents
- Share cards and gifts received from students
- The funniest thing a student has ever said

Elementary

- Video of an obstacle course in physical education
- Student success story
- Students reading the lunch menu
- A class play
- Fourth graders sharing fun facts about their state
- Share a Google presentation
- Read self-written poetry
- Fifth graders using iPads for a quiz game via Kahoot!
- Third graders sharing what they learned this past week
- Students helping students – mentoring/buddy programs

Middle/High School

- Video explaining their artwork
- Seniors telling of their plans for life after graduation
- Student speeches
- A trip to a local nursing home to play games with residents
- Band practice before an upcoming performance
- Math students giving a weekly math problem to your fans
- Foreign language students saying hi in that language
- HS students sharing the between-class routine
- Newly licensed drivers commit to not texting and driving
- Freshmen share their biggest fear about entering HS
- Interviews with foreign exchange students
- Drama practice followed by an invite to the performance
- Choir singing a Christmas carol near the holidays
- Seventh graders introducing the basketball team
- Juniors exploring the school swamp for biology class

Administration

- Superintendent's weekly shout-out to an outstanding student
- Principal sharing a teacher of the month
- Superintendent taking a video tour of a building project
- Special video announcement from the school board
- Weekly interview of the principal by a student
- Principal sharing answers to FAQs via video
- Video interview a student – What are you learning about?

- Take a photo of you having lunch with the kids
- Share the little things that make your school awesome
- Hide logo items around the school – fun for kids to find it!

Telling Our School Story

- Top ten reasons our school rocks
- Therapy dog in action
- Alumni feature – Where are they now?
- A day in the life of a student/teacher/principal
- Senior walk
- Teachers offer encouragement for test time
- Series sharing what kids like most about the school
- Parent/child reunited for military service
- Special volunteer invites
- Student of the week
- Parents expressing what they love most about your district
- Local businesses give a good luck message for the big game
- Highlight the impact a teacher has made

Events and Reminders

- Student invitation to a school fundraiser via a made-up song
- Student-coordinated blood drive
- A weekly preview of events
- Homecoming parade
- Video invite made by students
- Fun dress-up day reminders
- Community events supporting the school

Athletics

- Athlete of the week
- Score updates – use your school #hashtag!
- Challenge fans to wear their school gear to the game
- Thank the parents
- Practice photos or videos
- Video of victory song
- Coach's comments after the big game
- Fan of the week
- Athletes share what teamwork has taught them

Weekly Features

- #MotivationMonday – Well-known quote on a pic of your students
- #AlumniFealure – Where are they now?
- #TriviaTuesday – Ask questions around your PBIS curriculum
- #TransformationTuesday – Staff then and now
- #TipTuesday – A helpful tip for parents
- #WellnessWednesday – Share a favorite outdoor physical activity
- #WisdomWednesday – Share your best advice
- #TBT #ThrowbackThursday – Old yearbook photos
- #ThankfulThursday – What teacher are you most thankful for?
- #FlashbackFriday – Well-performing posts from the past six months

Summer

- Share an empty classroom, telling kids you miss them
- Summer projects, remodeling, and cleaning
- Recognize staff who work year-round
- Summer school
- Video update sharing information about back-to-school
- Display artwork from the past year
- Post lost-and-found items still at the school
- Ask fans what they are reading
- School supply list
- Next year's district calendar link
- First day of school reminder
- Search out your school hashtag and reshare
- Create a school pride contest
- Ask your community about their summer vacation bucket list
- Summer staff updates
- Feature students in their summer jobs

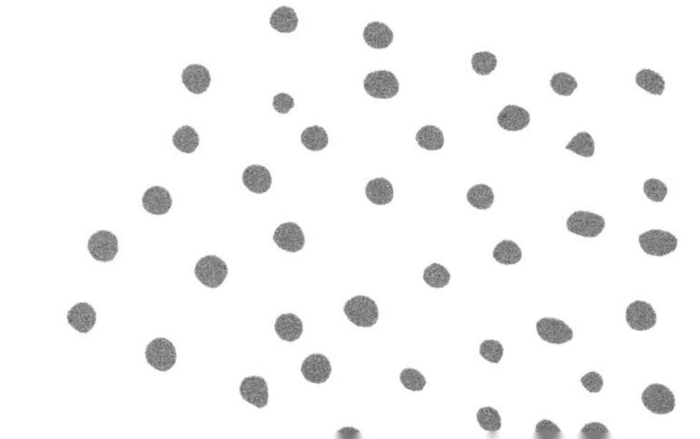

Grab it!

Grab a printable PDF titled **"100 Inspiring Ideas for School Social Media Posts"** with all 100 ideas from this chapter in the free bonus resources.

Plan Engaging Posts

As you post content, it's important to think about what creates engagement with the audience. You might be wondering why it matters to get engagement (likes, comments, and shares) on Facebook. Well, the algorithm favors content that engages users. The more your posts inspire people to like, comment, and share, the more likely Facebook will be to show that post – and your other posts – to your followers and their friends.[7]

The more your posts inspire people to like, comment, and share, the more likely Facebook will be to show that post.

You want to ask simple, engaging questions that are irresistible for your audience to answer. We've even compiled a list of fifty-two questions for you to use – one for every week of the year – so you can "set it and forget it" with a guaranteed post every week that will earn comments, inspire nostalgia, and connect your community.

Here are a few tips as you implement this strategy:

- ✔ Keep your questions to 130 characters or fewer. No long captions.
- ✔ Don't worry about creating branded graphics. Just create text-only posts using Facebook's preset colored backgrounds. (Facebook prioritizes these.)

- ✔ Consider how personal you get with your questions. We've seen schools receive pushback on questions like "what was your first car" and "what is the name of your first pet" because those are sometimes password security questions. You're not forcing anyone to participate, and you're not mining their data for nefarious purposes – so people can just be smart about what they post online!
- ✔ You'll see some of the questions are similar in nature, but they are going to earn different responses. When you space these out each week, they won't seem repetitive!

52 QUESTIONS

1. My favorite teacher ever was ________________.
2. Who was your kindergarten teacher?
3. Who was your first-grade teacher?
4. Can you name your second-grade teacher?
5. Who was your third-grade teacher?
6. Who was your fourth-grade teacher?
7. Can you name your fifth-grade teacher?
8. Who was your driver's ed teacher?
9. Can you name your favorite coach?
10. Name your elementary music teacher!
11. Who was your physical education teacher in elementary school?
12. Name your world language teacher. What did they teach?
13. Who was your favorite high school teacher?
14. Were you a band or choir kid?
15. Who was your bus driver?
16. What was your favorite school lunch?
17. What book was your favorite in middle school?
18. Who was your best friend in first grade?
19. What was the first organized sport/activity that you joined? How old were you when you started?

20. When you were little, what did you want to be when you grew up?
21. It's recess time in elementary school. What are you doing?
22. What was the coolest thing your school did that others didn't?
23. What was your favorite field trip?
24. What sports/extracurricular activities were you involved in?
25. What friends are you still in touch with today from high school?
26. What is your favorite season?
27. What instrument did you play in high school?
28. What was your favorite song in middle school?
29. What was your most memorable at-home project for school?
30. What volunteer opportunities have you been involved in?
31. Share your school colors using emojis!
32. You know you are from (insert town) if . . .
33. What was cool when you were young but isn't cool now?
34. Who made a difference in your life when you were a kid?
35. What would you like to tell your fifth-grade self?
36. What was your favorite game in physical education class?
37. What was your favorite subject?
38. What was your favorite school club?
39. Did you walk to school or ride the bus?
40. What was your favorite class in high school?
41. What was your favorite birthday treat?
42. What was the favorite thing you learned in high school?
43. What was your favorite homecoming activity?
44. Did you have a Trapper Keeper back in school?
45. If we looked into your high school locker, what would we see?
46. Drinking fountain or bubbler?
47. What do you call it: pop or soda?
48. What chores did you help with as a kid?
49. What was your favorite after-school snack?
50. What piece of advice would you like to give to your high school self?
51. What advice would you give to our graduates?
52. Did your school have a senior prank?

Let's try this now! Pick one of these and share it on Facebook. Then I want to know how it worked! Tweet me @andreagribble - or send me an email andrea@socialschool4edu.com.

Inspirational Quotes

I have one more tip for you as we wrap up this idea-packed chapter. For your #MotivationMonday or #ThoughtfulTuesday posts, you may want to share inspirational quotes from people. When you share these, note:

- ✔ You don't need stock images. Use your own photos with your students when you create the graphics.
- ✔ Choose quotes that represent diversity. Look beyond the traditional ones.
- ✔ Research the quotes to be sure they are accurately attributed. Quote Investigator is a helpful website for that. You might be surprised to know how many on Pinterest weren't said by the person quoted.

I hope you gathered some great ideas for your content plan!

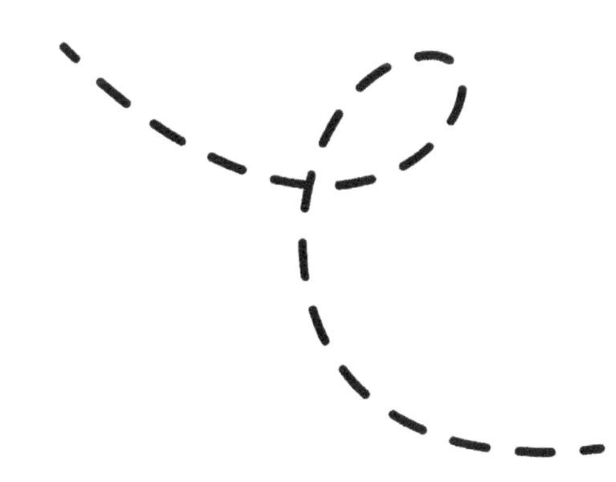

YOU'VE GOT THIS!

- ❑ Great content takes a village. Establish a system that works.
- ❑ Repurpose great content.
- ❑ Teach staff how to send you photos.
- ❑ Focus on engaging content with fun questions.
- ❑ Borrow ideas from our list of 100 great ones!
- ❑ Inspire with quotes from diverse sources.

What's your biggest takeaway so far in the book?

Tag me on Twitter at @andreagribble and let me know!

STORIES TO SHARE (AND WHAT TO SKIP)

11

Not everything needs to be shared on social media. Whew! That should be exciting news for you! If you must narrow it down to one priority, story is it. Think you don't have stories to share? Get ready to be inspired.

Case Study

Meet Nicole Valles: Nicole is the digital content specialist in Barrow County School System in Georgia, a district of 14,000 students. She's living proof that social media posts work for staff recruitment! She landed her role because of a Facebook post. Nicole is a former TV news producer with a background in journalism.

The challenge: As a former news producer, Nicole came from the school of thought that they had to get the communication out there. They had to own the content and be the source of truth. "While others are out in their parent groups spreading misinformation, we have to be the ones to make sure we are accurate and factual," she said. However, putting *all* communication on social channels led to some volatile conditions when people showed up ready to fight pandemic policies, which led to having to hide comments or shut off commenting.

The process: Nicole realized that communicating certain information via a different method wasn't hiding information. It was simply avoiding the invitation to publicly debate the decision that was already made.

The outcome: Not a single person complained about the change. Nicole said she has less stress now that she isn't following her honest-to-a-fault method on social media. She's no longer getting up in the morning with 1000 comments to moderate.

She isn't hiding comments and explaining the public forum guidelines daily. The change that was hard for her to embrace at first turned out to be something that she now loves.[1]

Stories are memorable. Stories create connections. Stories matter. That's why the best way to stand out on social media is to tell **stories** through your posts! And the best way to give your community stories they will remember is by highlighting students.[2]

What Stories Should You Find?

There are great moments happening all the time at school. Molly winning the spelling bee . . . Damien earning a scholarship . . . Kai getting a 100 percent on his math test.

Those moments are absolutely worth sharing, but I'm pushing you to find real, heart-warming stories to share.

Think of perseverance, determination, grit, or overcoming an obstacle. Which students have shown these characteristics? We want to tell their stories! And that story doesn't have to be a novel. Don't think that you need to write a long, elaborate story or that you need to create a lengthy, highly edited video.

How to Find Great Student Stories

The only way to discover the stories is to start looking! You should start with your staff. They have close relationships with the students and should be able to identify some ideas. Principals may be able to identify some students to highlight.

Attend the in-person staff meeting at the building level. If you can speak directly with staff, that may be the best way to ensure that they understand what you're looking for.

As a last resort, you can send out an email explaining what you're looking for. I give this as a last resort because you never know if your staff will read your email!

If you struggle to find options with staff, you could open your request to your community. This could be done via a social media post with a form included, allowing you to easily collect submissions from your followers.

Because of the sensitive nature of some of the stories, you may need to get approval from a parent or guardian. Don't skip this step! A simple phone call followed up with something in writing will be enough.

AARON'S STORY

TARA ADAMS, the communications coordinator for Lakeside School District in Arkansas, shared a highlight of a student named Aaron. Aaron was diagnosed with autism at a young age and is nonverbal. He also has a fighting spirit and loves the game of basketball. What a treat to see Aaron accept his position as student assistant for the LHS boys basketball team!

The caption with the post about Aaron said:

> We have so many shining stars on the Lakeside campus, but tonight we want to highlight Aaron [name redacted].
>
> Aaron was diagnosed with autism at a young age; He is nonverbal and continues to overcome obstacles. He has a fighting spirit and he loves the game of basketball.
>
> Recently, we were able to capture a special moment between Coach Lamb and Aaron. We loved seeing Aaron sign "thank you" to accept his position as Student Assistant for the LHS Boys Basketball team.
>
> We are so proud of Aaron and can't wait to see him on the floor this season. Go Rams!
>
> #WeAreLakeside[3]

The post included several photos of Aaron with the coach and teammates. It attracted more than 250 reactions, twenty-six comments, and nineteen shares.[4]

How to Share Student Stories

The first decision you should make after identifying your story is how you will share it.

- Will it be written, and you'll need photos to go along with it?
- Will you put together a video where the interviews will be shared?
- Will you do a live video to share the story?

Once you know how you'll share the story, you can conduct your interviews. I recommend direct interviews with the student, staff, or family involved. If you choose to do a video, you'll want to record your interviews. If you decide to write a story, you might still consider recording the interviews – at least the audio. Direct quotes will help your story come alive!

Check out "Examples of Great Student Stories" below for helpful examples to get your creative juices flowing! Note that images matter, as well as captions, when it comes to conveying your heartfelt story.

Student stories might include an adoption story, highlighting a special skill, celebrating finishing chemotherapy, and more. Challenge yourself to find those stories that deserve to be told. These features will positively impact your school and community, so they are worth the extra effort! If this sounds overwhelming, it really doesn't have to be. A great goal might be to find one of these stories per month to share on social media.

EXAMPLES OF GREAT STUDENT STORIES

Two years ago, Emi lost her lower arm and hand in a lawnmower accident. She has been on an incredible journey of strength but has been resistant to wearing "Ariel" (her prosthetic). A post from Oostburg School District in Wisconsin celebrated her perseverance.

This story reached over 10,000 people and had twenty-two comments and fifty-one shares! The photos, along with the short three paragraphs of text, are the perfect combination to tell a beautiful student story.

Meet Emi. She is a JK'er in Miss Rachel and Miss Brittany's classroom. Two years ago, she lost her lower arm and hand in a lawnmower accident. Her life changed dramatically. The last two years have been an incredible journey of strength and finding her own way, while staying resistant to wearing "Ariel" (her prosthetic) anywhere outside of occupational therapy.

School started and Emi told her mom she wanted to bring Ariel for when she played! This has brought many happy tears to her parents and family while also blessing our school community. Miss Rachel and Miss Brittany have learned a lot about Ariel and love looking for ways to help Emi and her classmates succeed.

They recently read the book *Different is Awesome* by Ryan Haack as they help their class see that Emi can do everything they can do, with some things done differently! Thanks to our JK team, Emi's classmates, and to Emi for showing how we can all learn from one another. How come the best lessons often come from 4 and 5 year olds? Way to go Emi!

#boosttheoost #lawnmowerawareness #limblossawareness #livingonehanded #differentisawesome[5]

To see the images and several other examples, see the link in the free bonus resources at socialschool4edu.com/book.

Skip the Information-Only Posts

Your social media page is not a bulletin board. Social media is the place to celebrate your school! We are living in a very divisive culture right now. I know that isn't a newsflash, but as a social media manager, you need to look out for the best interest of your school. As in, do NOT go looking for a fight.

If you know that new information about masks, or learning models, or school start times is going to invoke negative reactions, WHY would you post it on social media? I know, you might be thinking, "We need to be transparent. We need to get this information to everyone. We should communicate it any and every way we can."

Certain updates, changes in schedules, and modified procedures should be communicated through email, phone calls, or newsletters. Your staff may need some convincing on this because I know the default response for anything that needs to be communicated might be, "Put it on Facebook!" However, there are reasons not to share *everything.*

FIVE REASONS NOT TO SHARE EVERYTHING ON SOCIAL MEDIA

1. Your posts are not seen by all your followers. On Facebook, for example, if your school's page has 3,000 followers, the typical reach will be about 1,000 people. The reach will be less for posts with little engagement and more for posts with lots of comments and shares. But it's very typical to only reach one-third to one-half of your followers.

2. Not everyone is on social media. Many of your parents don't want to be, and if you are using it as your primary means of communication, they will miss it.

3. Outsiders with no connection to your school can comment on your posts. Do you really want to spend time dealing with someone's questions or opinions when they have no relationship with your district?

4. Social media is meant to be social. If you're simply making an announcement and don't desire commentary, then it shouldn't go on social media.

5. You're busy. Do you really have time to devote to monitoring comments, questions, and concerns that these posts stir up?[6]

USING STORIES FOR REMINDERS

How do you avoid info-only posts while still getting the reminders out there? **MELISSA HARTLEY** is the director of communications and marketing for Horizon Honors Schools in Phoenix, Arizona, a charter school district. She said once she joined the #SocialSchool4EDU membership program, she became more thoughtful about the way she was using her channels. "I really wasn't sharing news or celebration," she said. Instead of posting reminders on the main news feed, she now creates a Facebook event and shares reminders in Stories.

It's especially helpful as more people use Stories rather than scrolling their newsfeed. And the reminders only need to stay up for twenty-four hours.

Melissa pays attention to how she words her Stories because they might be visible after the event for a few hours. So, she adds specific dates to the text on Stories instead of saying "today" or "tonight."[7]

Grab It!

Your district has SO many ways to communicate. With each message that needs to get out, you need to strategically think about what tool is best to get the information out to the people who need it.

We have a thorough list with all the communication possibilities for you to use in your communication toolbox and decide which tools are best for which messages.

Look for the printable PDF of **"Your Communication Toolbox"** in the free bonus resources at socialschool4edu.com/book.

Special Celebrations and Holidays

Nothing creates anxiety in a school social media manager like the fear of missing an important national "day" on your Facebook page.[8]

Wait – what? It's Custodial Workers Recognition Day? And it's noon, and you have an afternoon of meetings . . . Darn it! You feel like a failure.

No, you're not a failure. I'm going to share the approach we follow at #SocialSchool4EDU in managing social media for more than eighty schools across the country. Our hope is that you'll have a plan for this school year to remove the anxiety, put you in the driver's seat, and ultimately feel positive about the work you're doing to celebrate your school!

Your first job is to celebrate your school. Every. Single. Day.

You don't need a special "day" to say that your principal is awesome.

You don't need a special "week" to give a shout-out to your bus drivers.

Now, those days and weeks exist, but if you're consistently celebrating great things in your school district, you should worry less about those national days. However, if you do observe them, you'll need a strategy.

A Strategy for Special Days

Since it can be tough to keep track of all the days and weeks out there, you should have a discussion with your leadership team about what days you want to recognize.[9]

One school in Minnesota shared with us that they were just stressed out from having to remember all the special recognition events throughout the year. Instead, they created an "Appreciation Month."

During this month, they take time to recognize all the different staff members that they have at the district – including support staff, volunteers, administrators, teachers – everyone!

This is a great approach! After all, I don't think that there is a national day to recognize paraprofessionals that help in the classroom (if I'm wrong, message me and let me know). I would consider them an extension of our teaching staff, but they don't carry the title.

This approach of having a month dedicated to appreciating everyone helps ensure you won't forget anyone! It's an option you might want to adopt.

Grab It!

National Toast Day? Take Your Chinchilla to Work Day? Which celebratory days should you include, and which can you leave out?

As a school communicator, part of your job is to celebrate those special days and help your staff and students feel appreciated – but sometimes, it can feel overwhelming! Our bonus PDF has a list of some of the most important days, weeks, and months to keep track of throughout the school year.

If you're interested in getting a detailed calendar for this school year, you might want to consider joining our membership program. All of our members get access to this each year.

The link to **"Calendar of Celebrations"** is in the free bonus resources at socialschool4edu.com/book.

If you choose to recognize the special days, weeks, and months on your social media channels, the biggest key is to plan ahead. You should look over the calendar at the start of your year in July and map things out. You can create your own editorial or content calendar by month. We have templates you can use each year in our membership group. Then at the start of each quarter, make decisions on how you'll recognize each event. Using photos of individuals or sharing how the staff members were celebrated on a certain day will perform better than just a graphic with "Happy School Psychology Week!"

Are you ready to get started with your action plan? Social media is perfect for positive stories. You have so much to celebrate among your students and staff. Focus on those stories. Let's build up your community in support of the school. Celebrate your school one story at a time!

YOU'VE GOT THIS!

- ❑ Make a list of the stories you know that could be shared right now.
- ❑ Reach out to staff to find out who has a story your community would love.
- ❑ Keep your ears open for celebrations and people who shine.
- ❑ Think twice before sharing information-only posts.
- ❑ Use all the communication channels you have – not only social media.
- ❑ Plan ahead for special days and holidays.

MANAGING CONTENT

12

Managing social media for your school district is a big task. If you feel you're falling behind and can never get ahead with great content, then you're in luck! With batching your work and these powerful content ideas, you'll be ready to manage your pages like a pro.

Case Study

Meet Amanda Keller: Amanda is the social media director for Weston School District in Wisconsin. She calls herself the accidental social media director because she didn't set out to serve in the role. She's doing amazing things in her small district of 300 students.

The challenge: Amanda was wildly successful with publishing Reels and getting attention but struggled with time management because of all the things she needed to keep track of.

The process: Putting some of the training that she received in the #SocialSchool4EDU membership group into practice, Amanda now batches posts that are regular features, posts such as #ThrowbackThursday and #FacultyFriday. She budgets her time, uses time blocks, and maps out her week ahead of time. Amanda also streamlined her process by using one email for gathering content.

The outcome: Amanda said her brain is always making lists of things she needs to do. "If I can, I brain dump as soon as I get to work," she said. She'll list out the things she needs to work on the next time she has a block and then puts them in a specific schedule. During her work blocks, Amanda said the structure allows her to keep working down the list.

Having content sent to that one email has helped her have content all in one place instead of

CASE STUDY

having to remember where it was – between texts, Messenger, and other sources that staff were using. As Amanda continues organizing her process, being a one-person department isn't as stressful as it used to be.[1]

If the idea of coming up with something to post every day sounds overwhelming to you, let me introduce you to some methods that will make it feel as if you pressed the *easy* button.

Productivity and *efficiency* are two buzzwords in today's working world. Everyone is trying to squeeze every ounce of production out of their to-do lists to keep up with the never-ending demands of their jobs.

School PR is no exception. In fact, I would argue that school communicators are under even greater pressure to handle more and more responsibilities within the same forty-hour (plus) workweek! So, I want to share one very specific approach that I believe will simplify your job!

It's called batching your work, and while you can certainly use this for many aspects of your job, I'm going to talk about batching your social media tasks and then go into more detail about *what* to include in that batched content.

What Is Batching?

Put simply, batching is picking one repetitive task in your work and completing as many repetitions as possible in a set amount of time.[2]

Cut down the need to post on the fly, and work ahead! There will always be last-minute events and quick-turnaround projects. But when you plan ahead as much as possible, you reduce that nagging worry that hangs over your head: "Shoot, have I posted on social media today?"

How to Use Batching in School PR

When you batch, you can choose to batch an entire process or just part of it. For example, #ThrowbackThursday posts typically perform really well for the schools I serve. You can take thirty minutes to snap as many yearbook photos as possible, then set aside thirty minutes another day to crop and sort them, and another thirty minutes to schedule them to post.

Alternatively, you could take enough yearbook photos to get you through one month and then sort, edit, and schedule them all at once.

No matter your approach, the result is the same: consistent, quality posts are ready to go on social media so you aren't scrambling at the last minute.

Besides #ThrowbackThursday posts, there are a ton of things you can batch. Here are some ideas:

- ✔ **Scheduling content.** If you have a submission system, you're probably getting emails throughout the day from colleagues sending in content. Acknowledge emails as they come in, but it's OK to set aside just one part of your day to actually schedule the posts. There's no need to post it the minute you receive content!

- ✔ **Choosing holidays.** There's practically a school-related holiday every day of the year. Scan through a master calendar and choose all the holidays you're going to celebrate this year, and make a plan. You can even send out some emails or schedule calendar reminders to get the content started.

- ✔ **Creating graphics.** Hop into Canva and create graphics for all upcoming semester events or #MotivationalMonday quotes. Another batched task could be creating standard graphics to get you through the school year, simplifying any last-minute posts that pop up.

- ✔ **Building features.** Social media features such as staff spotlights or senior shout-outs are easy to create in batches. Send out the survey first, then create the posts in batches as responses roll in.

- ✔ **Delegate projects.** You are just one person and can't be everywhere at once. If you just need to get designated photo snappers at upcoming events, spend an hour each month delegating to your content support team.

- ✔ **Finding quotes.** There are only so many ways to spice up a math class caption, so gather an arsenal of great education-related quotes to have at your disposal.

BATCHING TESTIMONY

Think of batching like meal prepping. Just like many families set aside a few hours every week to meal plan, grocery shop, and prep food for the week, school communicators can dedicate a set time to churn out social media content for days, weeks, or even months ahead. This will give you more time to focus on day-to-day tasks!

Consistency is the key component to celebrating your district. You will eliminate the worry of making sure your district is well represented, celebrated, and seen in the algorithms as well as ensuring you're keeping social media channels always active. Batching allows you to look, feel, and sound cohesive and consistent. [3]

—**STEPHANIE SINZ**, Chief People Office at #SocialSchool4EDU

Grab It!

We put together some resources to help you with your batching. Grab **"20 Inspirational Quotes for your School's Social Media"** and more from the free bonus resources.

Now that I've sold you on the benefits of batching, let's look at powerful features you might incorporate into your weekly content flow. There are many wonderful ways to celebrate your staff, students, and community members. Introducing new staff or posting staff shout-outs help people get to know your teachers, administrators, and support staff.

Student features showcase students, graduating seniors, and alumni, celebrating success and connecting generations in the community. And don't forget about featuring community members and businesses who have a tie to your school, volunteers, and family members. Some are alumni, and they support the programs at your school in so many ways.

You won't want to pick every idea to start at once – but grab a few that resonate with you.

Staff Features

In a guest post by Stephanie Sinz for #SocialSchool4EDU, she said, "Showcase your staff with an introduction that will allow followers to know them on a deeper level. Many of our districts schedule these posts for Friday mornings at 6:00 a.m."

To create this, Stephanie said to send an all-staff email asking everyone to submit the answers to the following questions:

- Name and position
- Career experience
- Family, hobbies, fun facts
- Why you love working at your school

She also asked them to submit one or two personal photos to go along with their feature. "You can also feature new staff when they start with your school. What better way to introduce and celebrate your new staff than social media? Introducing your new staff welcomes them into your district and community and allows followers to put faces with names," Stephanie said. [4]

SELECTING STAFF MEMBERS TO FEATURE

If you are in a large district, you may wonder how to select your staff members. I recommend rotating between buildings and choosing a variety of teaching and support positions. If you want to get simple recommendations of people to recognize, you could ask the principals to recommend two teachers and two support staff from their building. If you can select some of the popular faces that many recognize in a positive way in your community, you will really get some great engagement to kick things off!

Staff Shout-Outs

In her guest post, Stephanie Sinz said staff shout-outs "are quick, easy, and do not require any extra effort from your staff members. They highlight your staff like staff features but in a slightly different way. You'll still drive a ton of positive engagement!"[5]

How does it work? You rotate through staff and provide a simple post like this:

> Our #NewAuburn staff make a difference every day. Let's put our hands together for this week's staff member, Tina Schultz. [emojis]
>
> Thank you for all you do! [emojis]

You will watch the positive comments roll in!

Stephanie recommended connecting with your IT personnel to get access to staff school pictures and storing these photos in a shared folder. "Within the folder, separate the images by building, level, position, etc., to ensure you are featuring staff with even representation," she said.[6]

Using Canva or a design program like Photoshop, you can create a generic template to which you can add a staff photo.

You could also get nominations from your students, other staff, and the community to recognize the members of your school district in a staff shout-out. Create a simple form that can be shared on social media, in newsletters, and on the website. Let people know you are looking for stories of staff going above and beyond. Take the nominations and create a simple graphic including the staff member's photo. Each time you share a feature (maybe weekly or whenever you receive them), make sure to also link the form to receive more submissions.

Student Features

Social media is all about stories. And while your school is filled with hundreds or even thousands of students and staff who are collectively doing awesome things every day, sometimes we have to go back to the story of ONE.[7]

And that ONE story is the story of the student. These features do so much for your district! They highlight great students, of course. But they also help reflect the amazing staff you have in your school. These features build engagement because many times, people take the time to react or comment on the photo. Family members are also likely to share the feature, leading to an even bigger audience for your school.

Taking the time to highlight one specific student is a powerful way to connect with your community. It may be a story of kindness, compassion, accomplishment, or simply a celebration of the student. I'll break down a little guide to help you get organized.

When we developed a "Students Who Shine" weekly feature in New Auburn, Wisconsin, we first determined how students would be selected: that could be teacher nominations, peer nominations, or a committee that "knows" the accomplishments going on. We decided to ask each staff member to nominate one student who stands out from their peers. It could be for the extra effort they put forward (because school doesn't come easy for all students), for being a great friend, or for overcoming adversity.

If you already have a regular student recognition program like Student of the Week or Students of the Month, then make sure that you highlight those on social media! If it is a group of students, you could take individual posts to highlight each student one at a time. It may garner more engagement than a group photo.

Senior Features/Spotlight

In a guest post on different features, Stephanie Sinz talked about capturing senior stories. She suggested connecting with "your school counselor to learn about student plans for the future or create a Google form for seniors to complete."[8] It can be super short with two questions:

- ✔ What are your plans for the future?
- ✔ What words of wisdom would you share with younger students at your school?

Grab senior photos from the yearbook advisor. You don't even need to create a customized graphic. Those senior photos are beautiful, and their parents likely paid a lot of money for them. Most photographers give the rights to use them for school, so you shouldn't have any concerns with the right to use them. If you are unsure, you can talk to a few local photographers.

Other senior features could be more in-depth, Stephanie said. "You might include a short video for their post-graduation plans. Get creative! These students are your success stories, so don't miss out on celebrating them on social media."[9]

If you have 100 or more students in your senior class, this type of feature may be challenging to do. You could post four square photos at once on Facebook to fit them all in before graduation. Some schools choose to only post them on Instagram and not on Facebook.

Alumni Features

Former students are another source of community connection. If you want to get more alumni engagement on your social media pages, "start featuring them!" Stephanie said in her guest post. "You can initiate this content with a simple 'Calling All Alumni' post on social media or through your alumni newsletter." [10] Stephanie gave some examples of questions to ask alumni to answer:

- ✔ What year did you graduate?
- ✔ What did you pursue after attending our school?
- ✔ Who was your favorite educator, and why?
- ✔ What advice would you share with current students?[11]

You'll need a photo or two submitted from the alum – through your social media email address, of course! Stephanie suggested asking for one from graduation day and one from today.

Create a custom frame template, and you'll be able to swap out the images, add their name and graduation year, and schedule the posts.

Volunteers and Business Partners

Do you have a robust volunteer or business partner program? Make sure that you are giving the love to those people or businesses! You could plan to highlight at least one per month, and this could be captured in written story format or a short video. Your community will love to see their fellow members recognized, and it might help grow your programs for the school.

Families

Testimonials from families who choose to send their students to your school can work well for both public and private schools. There are often families who have had several generations attend your school, and highlighting these stories can be a powerful method to promote attending your school. You can select the best method (photo or video) and then aim to share one of these each month or quarter.

WHAT I LEARNED THIS WEEK FEATURE

"The big game. The flashy science experiment. The spelling bee. Graduation. All of these events are made for social media, right!?" Emily Rae Schutte said in a guest post for #SocialSchool4EDU.[12]

"But that's not what 99 percent of your school district is all about. Most of what makes your students and staff awesome happens in the classrooms," Emily said. Those stories from classrooms are even more important than the obvious posts.

Emily shared about Lakeville Area Schools in Minnesota that "found a way to crack open the doors to its classrooms and share everyday learning successes through the 'What I Learned This Week' feature." The idea was born out of asking how they could better highlight their students.

"Your feature could be a video, an audio clip, or a written quote," Emily said, but if you go with a written quote, don't forget to capture a photo of the child, too.

Stephen Rydberg, communications specialist for Lakeville Area

Schools, said that he and his colleague Grace Olson do not prompt or script the children they interview. "They simply ask the question and record the response. Of course, sometimes it takes asking a few times to really get a good response, but the key is to capture the authentic words of the child" in two to three minutes of filming, Emily said.

Before coming to film, "the Lakeville team lets the school principals know they are coming so that the principal can work with teachers to identify a couple of students to interview. They also track which schools and grade levels they are featuring, so they are able to get a good cross-representation of the district." [13]

Throwback Thursday Features

Last but not least when it comes to weekly features is #ThrowbackThursday, otherwise known as #TBT. If your school isn't yet sharing these fun blast-from-the-past photos on your social media pages, it should be. This is a great way to engage alumni and community members in reliving the old "glory days."[14]

Scanning the old yearbooks is a fun task. It's one of the things I look forward to the most when I visit a new school client.

If your school isn't yet sharing these fun blast-from-the-past photos on your social media pages, it should be.

Although it's fun, it can also be time-consuming if you don't know what you're looking for. You could quickly find yourself taking more than thirty snapshots from one single yearbook. That leaves just seventy more years to get through!

I limit myself to two hours, and I can usually get through eighty years' worth of yearbooks in that time, taking five to seven pictures from each yearbook. My advice is to take a photo of the yearbook cover first so that when you go back and organize your pictures, you'll know that the following photos are from that same year.

Once you have all the photos, save them in a TBT folder. Flip, crop, and rename them with the appropriate year, making it easy to schedule them out for every Thursday morning at 6:00 a.m.

Now, what should you look for when it comes to those images from each yearbook? I search for images showing:

- **School Spirit** – The face paint, the apparel, anything showing love for the school.
- **Technology** – It is so funny to see the old typewriters, movie projectors, and even computer monitors from 2000.
- **Memorable moments** – Think about the alumni that will see these posts. If it looks like there is probably a fun story behind the image, make sure to share that.
- **Big smiles** – Everyone loves to see happy faces – especially faces with eighties hair!
- **School pictures** – Grab a photo of a bunch of school pictures since these are always memorable!
- **Ads** – The ads that support the yearbooks are great to share, especially if some of those businesses are still in your community today.
- **LOL photos** – Whether it is the photo, the caption, or both, if something literally makes you laugh out loud while you look through the yearbook, snap a picture of it.
- **State Championships** – The entire community will love to think back to those memories.
- **Holidays** – Throwback photos of Christmas and Halloween can be great images to have on hand to share in those seasons.

Grab It!

Captions for your #TBT photos are important to drive engagement. We'll give you more of those ideas in the free bonus resources at socialschool4edu.com/book. Look for **"52 Weeks of #ThrowbackThursday Captions."**

Share the Love!

Stephanie Sinz said, "Seeing the engagement on feature posts is powerful, but have you ever thought of sending a personal note or email to the featured staff or student?" [15] What a great idea! Not everyone follows social media closely, and Stephanie gave two examples of ways to share the love with the people you feature.

First, you can email the featured person a screenshot of the post with the link attached. "Make sure to screenshot the comments," Stephanie said. That way, they can see all the encouraging input from your followers.

The second idea she shared was to print a copy of the post and description from Facebook. "You should include some comments and analytics as well," she said. A nice extra touch is asking the superintendent or principal to sign this hard copy and write a personal note you can deliver to them.[16]

Which of these features will you batch first? Once you see the positive energy the engagement brings to your page, I know you'll want to come back for more ideas.

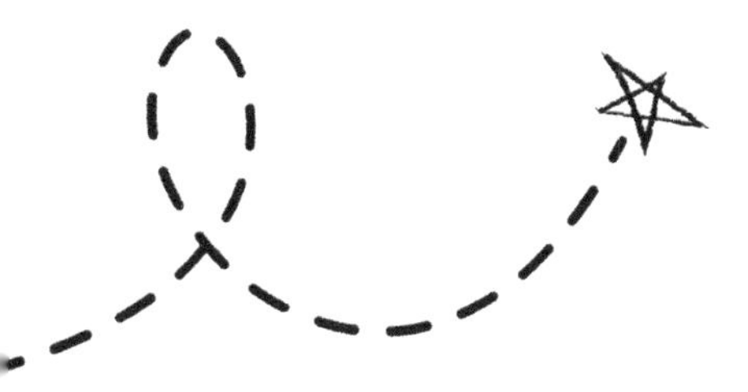

YOU'VE GOT THIS!

- [ ] Designate scheduled time to work on batching content planning, design, and scheduling posts.
- [ ] Store content in your cloud drive for easy access.
- [ ] Schedule posts on social platforms.
- [ ] Feature your wonderful staff, students, alumni, volunteers, and community members.
- [ ] Scan old yearbooks for fun, blast-from-the-past photos.
- [ ] Select one or two features that you will start doing on your social media accounts.
- [ ] Share the love by letting people know you featured them.

STAFF AND STUDENT TAKEOVERS

13

Whether you're a communications department of one or many, you don't have to do it alone! Students and staff can post, too. I'll coach you through how to loosen the reins and discover the win-win in takeovers.

Case Study

Meet Jason Wheeler: Jason is the director of communications for Garland Independent School District in Texas. He formerly served as director of communications in Carrollton-Farmers Branch Independent School District (CFBISD) in Texas, a district with thirty-eight campuses in six cities and 26,000 students. In a podcast episode of *Mastering Social Media for Schools*, he shared about student takeovers at CFBISD.

The challenge: Students have influence on their parents about where they choose to go to school and which programs they enroll in. But do they trust the voices of a district employee shouting out the accolades about the school and the great things you're doing? Jason said it's so much more powerful to hear other students' voices. But the biggest barrier is the perception that letting students have control of social media could lead to problems.

The process: Jason and his communications team implemented student takeovers and student-run accounts to get the message out directly from students. Kids love hearing from other kids. "Ease your fears," Jason said. Most students who share take pride in the school, and if someone does post something objectionable, it's easy to track who it was! The communications team developed a student ambassador program in addition to a parent one. Ambassadors don't have to be the valedictorian. There are "so many different people with so

CASE STUDY

many different passions," Jason said. An admin or a teacher serves as the gatekeeper and monitor as students create posts and Stories on their takeover day. His team used an older iPhone as a district-wide device to make takeovers on Instagram easier and bypass security and access snags.

The outcome: Giving students "the keys to the kingdom" boosted pride. Having students do takeovers surrounding events such as Homecoming increased engagement. Jason shared examples schools can implement: letting the band take over one day, the cheerleaders, the engineering team, or the robotics team on others. The program has increased teacher buy-in, but it's also grown followers. "Any time a student takes over," Jason said, "their friends are going to want to follow to see what they're doing." Their Instagram count went up by 575 people in one day because people wanted to see what the cheerleaders and dance team were going to do on Homecoming.[1]

SCAN QR CODE

FOR BONUS RESOURCES

socialschool4edu.com/book

Have you ever considered adding student contributors to your social media team? Perhaps you wonder how to get started or how other districts are utilizing students. Many districts use this strategy to increase student engagement and buy-in. Some schools are hesitant to let staff run their own pages or don't have the bandwidth to manage a whole process. Takeovers or story contributors are a great alternative.

You likely have a large following and instead of individual students or staff having to create their own social media channels to attract followers, they can jump in and boost your platforms!

Student Contributors for Social Media

Having students help might sound scary, but student contributors can be integral members of your social media team. Sara Fuller, an English teacher and social media supervisor in Superior, Nebraska, was a bit nervous about enlisting students. However, after a little training session with our team at #SocialSchool4EDU, she quickly saw the benefits. She said, "Students will be more themselves around other students which makes for more genuine posts. It's important to trust the students you've made part of your social media team."[2]

Sara suggested finding students who are passionate about telling your school's story. Naturally, journalism students are great at this. Student Council members, class presidents, and communications students all fit the bill.

Try to involve students who participate in a variety of activities! Once you've identified your contributors, make sure to train them. Have them follow other schools' social media accounts to get ideas. Discover their specific talents, such as photography, storytelling, and even those "extras" like emojis, filters, and GIFs.

Find students who are passionate about telling your school's story.

Some tips to keep in mind when adding student contributors:[3]

- ✔ Identify the right students.
- ✔ Spend time training. Show them how to log in, how to use a hashtag, what to do if they see something negative, and so forth.
- ✔ Have students sign a social media ambassador agreement.
- ✔ TRUST these students.
- ✔ Give regular feedback.

Sara's tip for involving other students is spot on: "We've done several day-in-the-life features for a volleyball player, football player, 1st grader, etc. and these Instagram Stories have been really popular." The fact that this school's Instagram account has more than 870 followers (with only 400 students K–12) is proof!

Student Tweeters

The New Auburn School District in Wisconsin has a new job for its class presidents. In addition to their other duties as leaders, each of these students has become a Twitter contributor on the district's account. These tweeters spread good news, updates in real time, and even staff shoutouts.[4]

After a short training session that included tips and tricks, these contributors were on their way. Accountability is key: students add their initials to each tweet so supervisors can keep track of posts.

Grab It!

Need more help? #SocialSchool4EDU has a free **Student Contributor Game Plan** that will help walk you through adding students to your social media team! The guide is linked in the free bonus resources at socialschool4edu.com/book.

STUDENT SOCIAL MEDIA AMBASSADOR PERSPECTIVE

"Instagram stories provide an amazing opportunity to reach more students and the community our district serves every year. It's also so fun to be able to keep everyone up-to-date on what is happening in the school and to see kids get excited about our social media pages and wanting to be involved! I've honestly seen an impact from the stories as far as student involvement goes, and kids are saying they know what's going on because of the updates they get!"

—**INGRID LYBERG** (from when she was a senior at Chippewa Falls High School in Wisconsin)[5]

Student Interns

When Amanda Oliver, Bellevue Public Schools director of communications in Nebraska, needed help, she decided to hire a graduating senior from the district to intern for the summer. In a guest post for #SocialSchool4EDU, LeAnne Bugay, the senior she hired, said the internship was "an instant success" for her and the district.

LeAnne explained how a student intern could be part of growing a dream social media plan.

> Planning and designing social media content can be a time-consuming task, especially if you're the only one in your department. But by hiring an intern and delegating that responsibility to them, you can grow your social media channels into what you've always imagined.
>
> In my experience, a large part of my role was to manage Bellevue Public Schools' social media channels. I spent a great deal of my time capturing candids inside the schools or designing custom graphics.
>
> Because this was my major focus, I could strategize our content for weeks in advance to create a storytelling powerhouse that the director had always wanted. We received several compliments from staff and parents about the new look.[6]

LeAnne said she was hired because of her work in the school's journalism department and her plans to pursue media in college. That "internship has elevated my skills in a short span of time," she said. It also helped her realize how much her district cares about the growth of its students.

That's a win-win! It's a way to help an overwhelmed district team member *and* support a student or graduate.

ADMIN PERSPECTIVE

ASHLEY MASON, former PreK–12 principal in New Auburn, shared her enthusiasm during her time at the school:

> As the principal of New Auburn, I truly believe that our school is the heartbeat of our community, and our community is eager to see what is happening within our school. The story we all are wanting to tell is about our students. What better way to have the story told, than by the people living it each and every day!
>
> Building up our students to be positive influences and leaders in our school is extremely important to me. I also believe that giving them this responsibility helps to foster that leadership they are all so capable of. I love having those moments in order to see our school day and what is going on in our school through their eyes.[7]

Grab It!

Listen to the full *Mastering Social Media for Schools* podcast episode for more details about the Studio 801 internship program and the internship application! Look for **"Engaging Students as Content Creators with Christine Paik"** in the bonus resources.

INTERNSHIPS BENEFIT COMMUNICATORS AND STUDENTS

In a podcast interview on *Mastering Social Media for Schools*, Christine Paik spoke about her secret superpower: student interns. Christine is the chief communications officer for Poway Unified School District, San Diego, California. In a district of 36,000 students with thirty-nine schools and a communications department of three people, tapping into student skills, talents, and passions has brought fantastic content creators to her team.

The Studio 701 Internship program is a prestigious program that includes an application and interview process. In addition to developing content creators, the internship also increases soft skills: professionalism, leadership, communication, and collaboration. From interviewing to writing, editing, filming, planning, and scheduling, these eight to ten students have skills that launch them into great careers. "A lot of our interns go on to major in either journalism, film . . . [or] PR," Christine said. They receive internship credit and a grade on their high school transcripts, which helps students in competitive college entrance processes, too.[8]

Staff Contributors

Not quite ready for students to take the reins on your social media accounts? Then how about starting with staff contributors? It would be best to select a staff member who is comfortable using social media personally. We have seen the best success with Instagram takeovers.

Staff could take over for a day to show your community a "day in the life." They could also help document a special event. If you wanted to get ongoing help from a staff member, you could ask them to share Instagram Stories as often as they want (since that won't screw up the posting algorithm on your feed).

When you arrange a takeover, you'll give short-term access to a staff member who posts photos and videos and interacts throughout the day.

You'll coach staff members ahead of their takeover, so you may wish to create a checklist, a short orientation video, or a toolkit to set them up for success. Some tips to include are:

- Examples of other takeovers
- Tips and best practices
- Ideas for eye-catching photos or video
- Caption ideas
- Tags or mentions
- Hashtags to include

When Ashley Mason was K-12 Principal at New Auburn, she used Instagram Stories several times per week to share what students and staff were up to in the district. One time, she shared a series of videos to show what the staff was working on during a late start day. The community got to see the various professional learning communities (PLCs) and they realized the importance of this staff-only time. Other times, she would share dress-up day photos or videos showcasing classroom learning. It was a behind-the-scenes look from her perspective on the district page. She didn't need her own separate channel; she was able to use the district page that had a great following to tell more great stories.

Instagram Takeover Posts and Stories

Richmond Public Schools, Virginia, has a helpful toolkit for staff takeovers. I've summarized some of their guidelines and ideas for posts here to give you inspiration for making your own toolkit. These ideas can easily be tweaked for student takeovers.

- Mandatory post: introduce yourself and your takeover for the day. This includes a photo of yourself, your family, or your work – not a graphic or clipart.

- Share about your career path, why you became a teacher, or how you got here and what you enjoy about your work. Include a throwback photo or a photo about what motivates you.
- Post with a favorite quote, a motivational book or movie, or inspiring hobby and encourage followers to share theirs.
- Share something lighthearted or fun that shows something beyond your teacher role: fun facts, a hidden talent, two truths and a lie, etc.
- Share something you are looking forward to, a goal you wish to accomplish, or a goal you have for your school.
- Who is someone you wish you could have dinner with? Did that person write a book? You could take a photo of it in your home or office.
- Share your favorite thing about your school or town.
- Share an exit post and thank the followers for listening, engaging, and joining.

Don't forget Stories in addition to posts! These can include mentioning other accounts, using hashtags, adding GIFs or stickers, and using music. You can add polls and ask questions or add a call to action. Richmond Public Schools also included tips for posting Stories throughout the day.

Here are a few ideas:

- Share photos of yourself or another introduction post.
- Share the original introduction post to Stories by clicking the "share" button once you've posted it to Instagram.
- Introduce your family, your pets, your team at school, your students, etc.
- Share any of your favorite things, hobbies, or school activities.
- Tag your favorite community organization, restaurant, or library.
- Post a final exit Story and thank followers for listening, engaging, and joining.[9]

The possibilities are endless, so encourage staff to be creative and have fun!

YOU'VE GOT THIS!

- ❑ Enlist the help of students and staff in creating and posting content.
- ❑ Establish a checklist or toolkit for guest contributors.
- ❑ Acknowledge the fear that comes with releasing some control and embrace the benefits.
- ❑ Celebrate how students gain valuable life and professional skills through working on your team.
- ❑ Appreciate the engagement that comes from other voices taking over for a day.

BEST PRACTICES

GREAT CAPTIONS

14

What if you're staring at a screen without a clue what to write? Don't panic! I've got your back.

Case Study

Meet Kristin Boyd Edwards: Kristin started her career as a newspaper reporter, then continued storytelling in her former work as a communications director for a school district in Pennsylvania. Her pro experience as a strategist and content writer keeps growing!

The challenge: Kristin did a social media training with a school where she demonstrated how to add a "little sprinkle of sugar" – as she calls it – to their captions. She pulled posts from their current Facebook page and walked through how to change them up a bit. One challenge was to make over this caption: "Does your school have a K-pop club? Ours does."

The process: Kristin said, "They were highlighting a fun school activity where the photos lack action." The group shots didn't have the detail to help the audience know what K-pop was, and they weren't going to resonate or increase engagement, especially if readers didn't know what the post was about. Sometimes the caption needs to be a little longer to resonate, Kristin said.

The outcome: "Move over BTS, our LGA K-pop club stars are taking center stage," she tossed out as a caption makeover. That pop culture reference to BTS would stop the scroll, but then she suggested adding a sentence to show people how it was a "really cool club, especially for middle schoolers

CASE STUDY

[that] allows members to learn about Korean pop culture and language through song dance and performances." She'd add a star glitter emoji, too, she said. "Anytime I can toss one of those in, too, it's a happy day." Kristin also recommended adding a link to an activities page on the school website to give a call to action: "See all of our extracurricular clubs."[1]

Using the same verbiage post after post doesn't drive the active engagement you are looking to promote. We've talked about finding your voice and getting inspiration from other sources, so let's pull it all together for your action plan and add some extra flavor to your captions.

Captions should be short and grab attention. Remember, your followers are often scrolling by on a mobile device, and your job is to get them to pause and read. Even more importantly, you want them to engage and react!

There really is no "perfect" length for a post. While short captions are often better, don't be afraid to add more context if it's needed! I always say that the first few sentences are the most important. If you have more to say, then write it out. Those who care will read it. Those who don't will get the main idea from your first two sentences.

Great Beginnings

Writing great captions starts with great openings. You could implement any of these with your next batch of posts.

In a guest post for #SocialSchool4EDU, Hannah Feller put together a list of tips with examples for how to grab fans' attention and get out of a posting rut.[2] Here are eight of Hannah's outstanding ideas and examples:

1. **Ask a question.** An easy way to switch up the tone of a post is by starting out with a rhetorical question. Try to keep things general and light-hearted; maybe you aren't even looking for a specific answer to

the question. Nevertheless, it makes someone reading it feel like part of your community.

> Who's ready to see some #LCatPride slam dunks tomorrow night?
>
> Aren't you excited to see how our #FallRiverPirates' Math 24 competition went?

2. **Use a colon.** One underrated way to connect two ideas is the colon. You can start out with a witty phrase and then stick a colon in to introduce the main topic of your post:

   > We have something to illuminate your Saturday: #NewAuburn blanket fort, pajama, and flashlight fun!
   >
   > What an amazing #LCatPride achievement: Lake Mills School District Exceeds Expectations!
   >
   > Cracking the #LCatPride code: students in Mr. Herman and Mr. Carroll's pre-calculus classes apply their knowledge of matrices and use cryptography to crack the code of an evil criminal mastermind.

3. **Find a fitting play on words or pun.** Who doesn't love a great play on words, especially for funny or light-hearted posts? Don't try too hard for this one, but if you mull over the topic for a second and a good pun pops into your head, go for it!

   > "Our #FallRiverPirates are flipping amazing! They were treated to a pancake breakfast by our awesome volunteers this past Friday."
   >
   > "#ReedsburgPride artists are focused on the "bigger picture," especially when they collaborate to create a mural in the halls of Webb Middle School."

4. **District hashtag or another experimental hashtag.** Putting your district's hashtag at the very start of a post is not only great for Twitter,

but it can also be an effective way to pop out in people's feeds. Use it as an adjective, noun, or other part of speech! Bonus points if you can make it alliterate with the word(s) that follow it.

> #LCatPride learners are up to something secret . . .
>
> #FallRiverPirates in Mrs. Doolittle's class have been working on a special project!

5. **Address a certain target group.** If you want to direct a certain post to a group of people, such as elementary school parents or alumni, why not call them out at the beginning of a post? That way, they know right off the bat that the message is directed at them.

> Hey, #LCatPride alumni! We need YOUR input on a special event we have coming up . . .
>
> Calling all #FallRiverPirates middle school parents: it's time to sign your child up for . . .

6. **Ellipsis (. . .).** Looking for a way to build suspense or lead up to big news? Ellipsis (dot, dot, dot) are your new best friend. They can jazz up a post and pair well with an exclamation point:

> The winners of this month's Good Character Award are . . . John Smith and Mary Warren!
>
> The location for the brand-new elementary school is . . . Hudson St!

7. **Alliteration.** It's so pleasing to the eye (and our inner voices, reading aloud) to see words in a row starting with the same letter. This will force you to think of creative, uncommon words and add a little polish to a basic post.

> #LCatPride Skating Saturday: Mrs. Zietlow spent time with a student and her friend who won her LMES PTO auction.
>
> Monday March Madness ensued in Mr. Brown's gym class!

8. **Introduce features consistently (using the same phrase).** It's important to realize that creativity and originality are great in posts, but using key phrases and words to signal certain types of posts is important, too! Phrases like "Staff Feature Friday" or "High School Athlete of the Month" right at the beginning of your post draw your fans to something familiar. This is crucial to creating a brand for your school, so it's OK to use the same short phrase every time for recurring features.

    STAFF FEATURE FRIDAY is back this week with another #LCatPride educator . . .

    It's time for the best day of the week: STAFF FEATURE FRIDAY!

Limit the use of all caps when posting on social media. This is due to ADA compliance. Reading tools designed to assist those with vision impairments will read each letter instead of the word. You should also limit the use of specialized fonts. These, too, can have issues being read correctly.

Hannah said, "Hopefully this gives you a little push to vary the way you start posts! The first step is being conscious of the way you word things; once you start paying close attention, you will become a pro at creative opening phrases."[3]

Grab It!

Need inspiration? Check out our lists of schools with award-winning social media. Follow each of the schools listed on your social media platform(s) of choice. Screenshot the post ideas you like the best and save them in your "ideas" folder for the next time you need some post inspiration. Look for **"Schools with Award-Winning Social Media to Follow"** and **"Caption Clinic: How to Stand Out on Social Media with Creative Writing"** in the bonus resources.

FINDING YOUR VOICE

Find your voice. And this will come in time as you begin to build your social media presence. I always use *TMZ* and *Good Morning America* as examples. They could be sharing the exact same information or news story, but their posts sound different. And not saying that one is better than the other, but each news outlet understands their audience, and so they've cultivated an authentic voice that engages their readers. I just encourage you to do the same.

One of the examples that I use for this is in Pennsylvania, and I'm sure in many other states we have standardized tests. Each time Reading would grow our test scores, I would use rap lyrics in our social media post. . . . That is something I knew our families would connect with, and it would get them to stop the scroll because they would see the rap lyric and want to know what we were talking about. [4]

—**KRISTIN BOYD EDWARD**s, writer, strategist, content creator

ADD SOME FLAVOR

There are some tools to help you include more than plain text in your captions. Of course, special effects are best when used in moderation![5]

- ✔ Hashtags – Using hashtags can draw more eyes to your posts! We covered those in chapter 7.
- ✔ Bullet points – Try using bullets to make your posts more readable. You can copy and paste many fun bullet options from a link in the bonus resources.
- ✔ White space – A lot of dense text is difficult to read. Your brain needs to see space on the page! As a general rule, don't have more than three or four lines of text together without breaking it up into paragraphs with space between. And we are talking about three to four lines on a mobile device, not the computer screen!

Grammar and Style

Before you type out a caption and hit post or copy and paste from an email, pause to polish it up. Is the post in your voice? We covered that in chapter 6. Is it engaging? Does it have info you don't need? Have you checked for errors?

In several guest posts for #SocialSchool4EDU, Hannah Feller provided excellent advice on making sure the captions are outstanding. Here are a few of Hannah's tips:

- Edit out unnecessary dates. If the date a teacher sent with a post distracts from the message, trim it out.
- Post on behalf of the district rather than a single class. Change "my class" to "Mrs. Brown's class," for example.
- Drop in the hashtag.
- Change up your word choice to avoid being repetitive.

Hannah's example of a caption makeover:

BEFORE

Subject: Science Experiment

Today, March 15th, 4th graders in my class did a science experiment involving baking soda, vinegar and food coloring. We had so much fun and can't wait to talk more about chemical reactions tomorrow. Attached you will see pictures from the experiment.

Thank you,
Mrs. Brown

AFTER

#FallRiverPirates science enthusiasts "bubbled up" with excitement after witnessing a chemical reaction firsthand! Mrs. Brown's fourth-grade researchers executed a riveting experiment with just a few common household items: vinegar, baking soda, and food coloring. They cannot wait for more #ChemicalReactionTime.[6]

Here are a few words of advice inspired by Hannah's guest post:

- Understand what needs to be capitalized. For example, school district is capitalized when you talk about Fall River School District, the proper name of the district. However, when referring to the school district without the full name, it's treated as a common noun. Unnecessary use of capitalization looks less professional.

- Use the exclamations sparingly. Yes, they convey excitement, and we want to be positive and upbeat in our schools. However, using them for every sentence weakens all your sentences. Choose the most intriguing sentence in your post and use an exclamation point for that one. If you need two, try not to use them in back-to-back sentences.
- Learn where to put commas. You'll find some excellent tutorials in grammar guides.
- Say it in as few words as possible. Be clear and concise to stop the scroll.[7]

Writing skills take time to polish, and mistakes happen – even with great tools (my team uses Grammarly). Spelling and grammar checkers can't distinguish some errors. Do your best, give yourself grace when needed, and learn from your mistakes.

THE VALUE OF SLOWING DOWN – AND USING SPELL CHECK

In a district she used to work in, **ERIN MCCANN** was operating solo as the district communicator when she discovered the challenge of having many eyes on every post on social media. She was at a conference and received a call from a principal asking for her to create a release. "So, I stepped out and wrote it, didn't take a close enough look at it, and sent it off. . . . Luckily, it ended up being a really silly mistake. . . . It was the wrong grammatical word, so it wasn't anything about crushing anyone's reputation. But man, that community that day just decided to go to town on my grammatical mistake," Erin said. She admitted she knew she should have asked someone to look at it or run it through a grammar checker. "A huge learning point for me is that when I get stressed, I move too quickly, and when I move too quickly, that's when I make mistakes."[8]

HOW TO MAKE YOUR SOCIAL MEDIA POSTS REALLY "POP"

Ready to take your posts from bland to bam? Think about what *you* really like seeing on social media, and are most likely to like, comment on, or share. Here's a quick way to inject some energy into your caption: Try to **teach**, **entertain,** or **inspire** the reader. Here are some examples Kristin Boyd Edwards shared in a webinar for #SocialSchool4EDU members:

» Find the emotion – and tug at the reader's heartstrings. Nostalgia, accomplishment, triumph, overcoming, celebration, feel-good, high-fives, awws!

» Ramp up emotion and tell readers why they should care.

» Give a behind-the-scenes glimpse.

» Show your vibe and what's special about you.

» Teach readers and share useful information.

» Connect the dots: if this, then that.

» Get the reader involved with questions and calls-to-action.

» Entertain – make people laugh!

» Provide nostalgia or jog my memory.

» Show an action: a classroom project, experiment, learning in action, kindness on display, fun dress-up day.[9]

Emojis

Stephanie Sinz, chief people officer at #SocialSchool4EDU, said, "Emojis are a great tool to have in your back pocket. Consider the possibilities for using emojis and the power they bring for branding and communication. Using emojis in social media posts can enhance your caption, provide a breath of fresh air, and attract followers to the content you are sharing."[10]

Consider the possibilities for using emojis and the power they bring for branding and communication.

With over 3000 options for emojis in existence, there is no end to the variety they can offer to your captions! Stephanie said one study found that using them in a Facebook post can greatly expand the number of likes and comments on the post.[11]

Here are several of Stephanie's ideas for how to incorporate them into your captions:

Science experiments – Adding an equation of emojis can be powerful to enhance any science experiment happening within the walls of your district.

Hearts – Utilizing a simple heart that represents your school colors can help spread the love in your caption. Positivity is infectious!

Thumbs up or hand symbols – Give your students and staff a thumbs up or high five to illustrate how proud you are of them. And encourage followers to interact by adding their own thumbs up in the comments.

Music concerts and performances – Add music notes in your concert posts.

Seasonal – Snow, flowers, and sun can accentuate the time of year in posts.

Animals – When a class makes a trip to a farm or a zoo, animal emojis make a fun addition.

Honors medals – Highlight student accomplishments with an emoji medal.

Spice up your voice – Catchy phrases such as "Have a meowy Monday" or "The cat's out of the bag" are extra cute with a fitting emoji.

Universal symbols – The symbol for recycling is perfect on a post about a cleanup for Earth Day.

Sports – Whether it be running, softball, baseball, gymnastics, bowling . . . there is a large group of sports symbols from which to choose.

Food – Need we say more? There are a ton.

Flags – Celebrate connections with other countries.[12]

Need help finding an emoji? I like emojipedia.com - I can type in a word or phrase, and it helps find them. There are also keyboard shortcuts (For PC users: Windows + Period; For Mac users: Ctrl + Cmd + Space), and you can search for emojis, as well. For example, you could type in purple if you would like a purple emoji.

Calls to Action

We've discussed calls to action a few times already, but you can never have too many ideas. Here are a few you can personalize to your voice. If you want people to engage with your content, you have to ASK them to do it.

- Give our #(school hashtag) students a thumbs up!
- Give this post some love if this made you smile.
- Give this post some love if you agree that our #(school hashtag) students are __ !!
- We couldn't be prouder! Give this post some love if you agree.

- Found this motivating? Send it to a friend!
- Share this to help us spread the word!
- If you enjoyed this, share it.
- Show your #(school hashtag) pride – share this post!
- Save this post so you can come back to it!
- Tap that "save" button so you can come back to this later.
- Screenshot this for later!
- Make sure you screenshot this for later.
- Drop an emoji below if you're feelin' this.
- Tag someone who would love to see this.
- Let's give __ a big_ [applause emoji] applause in the comments!
- Drop an emoji below if you're as proud of __ as we are.
- Visit our website to learn more.
- Intrigued? Visit the school website for all the details.
- Check out the link below for more info.
- Check out the school website for more information on __ .[13]

Grab It!

Grab a printable sheet with all twenty calls to action shared above from our website. You'll find **"20 Calls to Action to Drive engagement on Social Media"** in the bonus resources at socialschool4edu.com/book.

YOU'VE GOT THIS!

- ❑ Get the attention of your fans with creative text, bullet points, alliteration, or puns.
- ❑ Tap into emotion with your brief, descriptive posts.
- ❑ Ask questions and include a call to action early in your caption.
- ❑ Use emojis that fit the mood and occasion of your posts.
- ❑ Break up longer text into shorter sections.
- ❑ Proofread, double-check spelling and grammar, and polish before posting!

HANDLING NEGATIVE COMMENTS

15

Don't let a few incidents of negativity discourage you from telling the amazing stories of your students. With a solid plan, you'll have no problem managing a stray negative comment here and there.

Case Study

Meet Holly McCaw, APR: Holly is the director of communications for Otsego Public Schools. It's her twelfth year in the rural district on the west side of Michigan with 2400 students. There are three elementary schools, one middle school, and one high school. The district has a website and district-wide social media pages.

The challenge: Holly has dealt with her share of negative comments on social media, but she sees them as an opportunity, not just a problem. So how does she handle them?

The process: Holly remembers some advice she received when she was new to school PR: if one person puts it out there, others might be thinking it, too. So, it's an opportunity to steer the conversation and provide correct information. Holly thanks the person for commenting. She then provides a source for more information and invites the person to reach out to her if they would like to talk more. Her district keeps public guidelines on its communication page, so if there is a personal attack or name-calling, she can go back to the rules to refer someone to the expectations regarding commenting, and it empowers her to block people if needed.

The outcome: Holly's actions have changed as social media evolves. Some policies have needed to be updated. Some comments that wouldn't have given pause in the past might create a red flag now

CASE STUDY

because she knows where the audience is at with current events. The district has changed its policies over time about posting certain announcements on social media because they have the potential to draw negative comments. COVID announcements fall in this category.

Most feedback is positive, and handling negative comments is rare. But Holly is prepared to act as needed. She said, "You're not going to please everyone, so don't get too concerned about [some] negativity coming your way." Even in the worst challenges, she has found that it's only a few voices making the most noise. She also said, "Take a step back and see how big of a deal it is. Is this one person giving their opinion? Maybe it doesn't even need to be addressed because people are allowed to have their opinions. It might not be what you want to hear, but they are allowed to have their opinions, so you can't fire back on every little thing." Sometimes a watchful approach is all that's needed.

Handling negative comments has also had another effect. Holly said, "It has made us all better writers. Maybe a little bit more strategic." As followers increase, she continues to be strategic about telling great stories and staying true to the purpose for social media.[1]

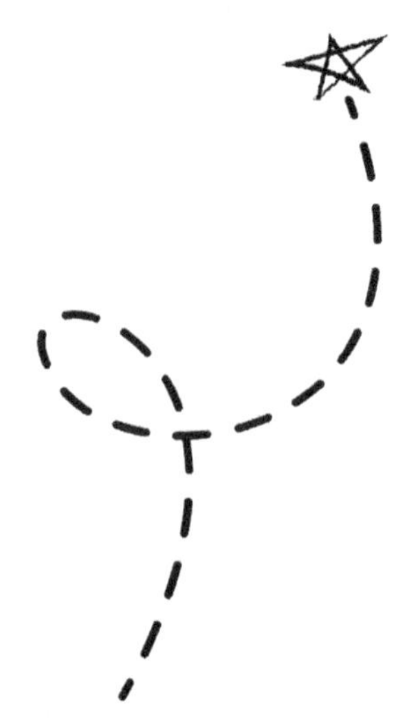

It's every school's biggest fear. Someone takes their dissatisfaction with the school to a whole new level by bashing the district on your school's Facebook page. Or perhaps they voice concerns on a private community page.[2]

After millions of posts and serving well over 100 school districts, I can honestly say that 95 percent of the time, the comments are positive. The amazing stories of your students put a smile on so many faces.

But with that said, it is a valid worry. You may be really stressed out if you don't have a written policy for handling these issues. Rather than waiting until something happens, schools should create a plan for dealing with these instances before they arise.

Negative Comments on the School Page

Be prepared for comments that you have to deal with. Here are five strategies to have in place ahead of time to help you navigate if negative feedback happens.

1. Document a complete flowchart on how to respond. That flow chart can contain problem-solving questions to direct your next step.[3]

Examples of a possible flow:

- Is the site credible with many viewers?
    - » Monitor only: avoid responding to specific posts but monitor the site for relevant information and comments. Notify your supervisor.
- Is the posting a rant, rage, joke, or satirical in nature?
    - » Monitor only: avoid responding to specific posts but monitor the site for relevant information and comments. Notify your supervisor.

- Is the post the result of a negative experience?
  - » Talk to your supervisor about possibly contacting the person who made the post. After a personal conversation, consider posting a clarification.

Grab It!

You'll find a flow chart example called **"Guide to Responding Online"** in the free bonus resources at socialschool4edu.com/book. Use this for inspiration for creating your own response plan.

2. Ensure someone is monitoring the comments. Social media is not a 9:00 to 5:00 (or 7:30 to 3:30) job. Someone on your team should be watching the interaction during non-school hours, or negativity and false information could escalate quickly. Social media platforms have settings to automatically alert you when there is activity on your site. I recommend that you check in several times a day on your platforms. You don't need to be on 24/7, but you need standard check-ins, even on the weekends. And, of course, if you anticipate negativity for any reason, be extra cautious.

3. Develop a standard response that can be followed. Thanking the person for bringing the issue to your attention is the first step. Then you can follow it up with an invitation to discuss the issue in more detail offline. #SocialSchool4EDU often shares this example of a complaint regarding school being canceled due to weather.

> Thanks for bringing up your concern regarding school being canceled today. The safety of our students and staff is one of our highest priorities at ABC School District. The decision to cancel school today was based on several factors, and we would be happy to discuss these if you'd like to contact the administrative office at 555-5555.

Remember that social media commenting policy you created back in chapter 2? This is where you need to put it to use. Abiding by the rules you laid out in that policy, you can make decisions on what comments you respond to, what you might hide, and what you can delete. I don't recommend deleting comments without getting your legal team and administration involved due to open records requests. If you delete it, always take a screenshot of the comment and save it in a folder first.

4. Don't get into a war of words on social media. And make sure your staff doesn't, either. It can be tempting to call out misinformation when you know the other side of the story, but there can often be privacy matters for students or staff involved.

5. Fill the social media channels with positive stories! The best way to move past a negative comment is to fill up your social media channels with amazing students, fun videos, and staff stories that will yield positive feedback.

Social media is meant to be social. When you receive comments, whether they are positive or negative, this is a great thing for your school. Opening up the conversation and letting people be heard is important.

The best way to move past a negative comment is to fill up your social media channels with amazing students, fun videos, and staff stories that will yield positive feedback.

Can I Just Turn Off Comments?

Can you relate to this concern from one of the school communicators in my membership group? "My district likes to turn comments off on posts because they worry about students or others saying bad things. What should I do?"

I don't recommend turning off comments, and I'll explain why in a minute, but ultimately, if that is what your administration has dictated, then you are going to have to follow their rules. I realize my entire book is sharing best practices and strategies to maximize your social media efforts, but you might not have the authority to start doing or trying everything. I think the best thing you can attempt to do, if this is your school, is rationalize the reasons why you don't want to shut off commenting.

On both Facebook and Instagram, you as the page manager can limit comments on a post-by-post basis, and YouTube also allows comments to be switched off for the entire channel. However, it isn't recommended to turn off comments completely, as it can sometimes cause more issues with your audience.

Because of the algorithm, if you disable or limit comments, your posts will automatically reach fewer people. This is because there will be less engagement on the post, and less engagement means fewer eyeballs because the platform doesn't think it's worth it.

Your fans will also realize that comments are restricted. If you want to poke the bear, keep putting out content that people can't respond to. Remember – social media is meant to be social. If you don't want feedback, you shouldn't post it.

Many of the social platforms provide methods to moderate content. You can hide profanity and other specific words from appearing on your page. In some cases, you can even hide comments with links included and posts from accounts with no profile photo (which assumes it might be a fake or bot account).

With some of this logic, you may be able to get your leadership to change their minds about allowing comments. You could at least ask for a trial period to see how it goes. Many people think the worst but experience the best when it comes to all the positive feedback they receive about students, staff, and the school!

Grab It!

I wrote a lot about comments in a blog post titled **"Controlling Comments on Facebook."** That help is available for you in the bonus resources.

BEFORE YOU RESPOND

Establish guidelines for your colleagues who want to respond to problematic comments on social media. The first thing they should do is contact a school leader or communications professional. And in all cases, remember these tips when you respond, which should be included in your flowchart.

- ✔ Be transparent by stating your connection to the school.
- ✔ Cite your sources by using hyperlinks, video, images, or other references.
- ✔ Take your time to think through your response. Don't rush.
- ✔ Respond in a tone that reflects the professionalism, empathy, and mission of your school.
- ✔ Give thought to the most effective type of response. Don't debate an issue or get into a negative conversation online.
- ✔ If appropriate, contact the commenter privately for a resolution and follow up online with an update or apology.

If you have something happening where you anticipate a lot of disgruntled families, you may choose to "go dark" on your social channels. That means not posting anything new for a day or longer. We've navigated employee arrests, student or staff suicide, and board decisions that are not favored by half of the community. In those times, we stop posting for a few days so the negative chatter can happen on news sites. Once we are ready to start again, we go slow. Please note that it is very important to look at scheduled posts in times like these. You wouldn't want to post a photo involving students or staff who have been involved in the situation you are concerned about.

Community Gossip Groups

I've received an email message like this more than once. Can you relate?

> Andrea, two of my school board members want me to join the gossip group in our community on Facebook. They expect me to respond to every post that discusses the school. And the only real way I can join the group is with my personal profile. What should I do?

You can't control what goes on in community groups. But you also can't pretend that they don't exist. Here are six strategies you can implement.[4]

1. Be aware, but don't join.

Identifying online community groups that like to talk about the school is important. It's like being aware of where the quicksand is located on the trail that you're hiking.[5] (Seriously, as a kid, I really thought quicksand was going to be a bigger issue!)

But just like quicksand, you should intentionally avoid being put in harm's way. While it's certainly tempting to join these groups so you know what's going on, trust me when I say that it's *way* too easy to get sucked in.

Because most of these groups are set up to not allow Facebook Pages to join them, your only option is to join as your personal profile. Before long, you'll spend more time than necessary reading comments and jumping into discussions. Pretty soon, people will start to tag you anytime a school question or issue comes up. They will think that the Facebook group is their direct line to all things school related.

In other words, you'll always find yourself in reactive mode rather than proactive mode. Is that really something you want to have to manage on top of every other responsibility? My gut reaction is NO. You don't have time, and you certainly don't get paid enough!

If you do join, listen, but don't necessarily respond. You are not responsible for responding to every comment about your school on other pages. Let's face it; you'd definitely run out of time in your day.[6]

You are not responsible for responding to every comment about your school on other pages.

Besides, you are rarely going to change someone's opinion by commenting on social media or an article online.

You don't want to assume that everyone in your district feels the same way as the vocal few.

Try to think of your community like the audience in *The Muppets*. They are diverse and likely an entertaining, lively group. That is the reality you have to learn to deal with in a constructive manner.

But when it comes to negativity, sometimes the only voices you seem to hear are the people criticizing your school.

If you start communicating as if you are only responding to those critics in the balcony, you'll veer down the wrong path. The others in your community will wonder what you're even talking about! At times, facts may need to be highlighted in your normal communication channels, but you certainly don't need to react to every comment out there.

2. Build relationships with advocates.

Instead of joining the group(s) yourself, find a few trusted people you know that are in the group(s) and who can let you know if there is a hot topic related to your school. Remember, they're there to be your eyes and ears, not your voice! They should simply relay information back to you as the communications pro to address in your own way, as needed.[7]

Why? Many parent-run pages will eventually self-correct. As Kristen Sutek, the community relations manager for Pine Lake Prep in North Carolina, said, "A reasonable parent will show up to shut the circus down, but it might not happen quickly enough."[8]

That's where your listeners can also become your advocates – if they are comfortable! These helpful, positive parents or community members will look for opportunities to steer people in the right direction.

An easy way to help these advocates is by giving them a simple script to use in their comments, such as:

> "I think the best place to find that information would be the school website: ____"
>
> or
>
> "This sounds like something the school would talk to you about directly. Have you tried reaching out? Here's the phone number: ____"

These gentle redirects may shut down discussions before they spiral out of control. Remember, you're not asking that person to defend every decision the school makes or to be able to fully explain the school's position, but rather to be a voice of reason!

3. Move discussions offline.

Perhaps there are a few persistent voices of dissent, negativity, or misinformation in your community group. It could make sense to start some offline conversations.[9]

Todd Porter, director of communications for Jackson Local School District in Ohio, shared this: "We actually invited the moderator of the group in for a forty-five-minute discussion with the superintendent and other members of the admin team during the summer. This makes them feel like they are getting a look behind the curtain. More importantly, they feel as though they are part of the team and therefore part of the solution and helping us mitigate rumors and bad information."[10]

A quick phone call could also do the trick, especially if the person is known to your school – such as a parent.

POSITIVE PARENT GROUP

JACKIE BRAUSER, the director of admissions and development for St. Francis Xavier School in Ohio, explained how a group of parents helps with communication:

> We use our Positive Parent Group to submit Google and Niche reviews, consistently share and interact with our social media posts, and help with testimonies. The parents also try to corral a conversation if it gets out-of-line on our school community (unofficial) social media page. We are a private school, and this group created last year (2021) is a godsend. If they can't handle a comment or post that a parent has made, they report it to me (I usually already know about it, but I am silent) to take the appropriate next steps.[11]

4. Correct misinformation through your own channels.

If some rumblings just won't go away, it could be worth addressing the gossip head-on through mass communications. I recommend direct communication with families whenever possible.[12]

For example, let's say community groups are sharing incorrect information about your school's new dress code policies. And despite your advocate's efforts to point them in the right direction, questions and confusion still circulate.

In this situation, you could send out a direct communication to all families that clearly states the school's policies.

You could also choose to correct misinformation on the school's official social media page(s). Cala Smoldt, communications coordinator for Sherrard CUSD 200 in Illinois, shared this advice: "Know what's happening in those groups (but don't obsess). If there are big ongoing conversations or misconceptions – it's an opportunity to put out information from the school with the correct info, in a subtle way. For instance, if we see something going around, we will address it in our 'tiger talks' (where the superintendent talks somewhat candidly about what's going on across the district, need-to-know info, etc.). It's embedded in the info, but it isn't the main focus."[13]

5. Continue celebrating your school.

When you are under attack, the best thing you can do is to keep telling positive stories about your school through social media channels.

Double down on the optimistic stories that you are putting out. If you aren't posting at least two times a day about the great things going on in your district, make that your goal.

One Minnesota superintendent said it best: "You have to build a culture where it is not socially acceptable to attack our school." It certainly takes time, but it can be done. You can build a place where positivity is the norm and where everyone works to keep it that way.[14]

6. Encourage your community to get involved.

There is power in numbers, and the more people sharing your positive stories, the better! The use of your hashtag will make it easier to see all the great things people share.

Asking for help in telling your story can be scary. It requires placing trust in other people to share the good things that are happening in your district. But in my years of studying social media in schools, I have witnessed a multitude of benefits and rewards when leaders trust their community to help them tell their stories!

....

I hope this empowers you to draw a hard line regarding your presence – or the pressure to have a presence – in community gossip groups.

Remember, most comments are positive.

YOU'VE GOT THIS!

- ❑ Always have someone monitoring comments.
- ❑ Pause before jumping in to defend your school.
- ❑ Create a flowchart with a clear action plan for how to handle negative comments.
- ❑ Counteract negative comments by posting more positive content and amazing stories.
- ❑ Be proactive rather than reactive.
- ❑ Find advocates to be eyes and ears to alert you to any buzz about the school on community pages.
- ❑ Correct misinformation through your own channels rather than public debate.

VIDEO TIPS

16

Real students and teachers telling their own stories on video increases engagement and leaves an impact much greater than words or images alone. But who has time? It's easier than you think!

Case Study

Meet Laraine Weschler: Laraine is the communication specialist for Naugatuck Public Schools in Connecticut. It's a district of 4300 students in a larger town that serves a diverse population of students.

The challenge: Short-form videos gain traction on social media, but videos take more time than photos. How does a communicator make that work?

The process: Laraine saves the fully produced videos with B-roll, voiceovers, and other effects for special occasions or purposes such as staff recruitment. But for everyday posts, she doesn't need to invest much time. She adds a few clips with music, keeping popular trends in mind. She urges others not to overthink it when it comes to creating video. Followers don't expect a professional-quality video, she said. Although Laraine has a "fancy camera" that can take quality videos, she said the extra time to edit those isn't worth it to her. She wants to quickly put up a fifteen-second clip of the sports team winning the semi-final match.

The outcome: The district has seen followers slowly increase. Laraine looks at what gets the most reactions to influence content she'll create in the future; she keeps an ongoing list of ideas for future video to capture. "Sometimes a video can tell you more

CASE STUDY

than a picture," Laraine said. When telling the story of student accomplishments on video, "you can hear the enthusiasm in somebody's voice" or hear people cheering in the background, she said.[1]

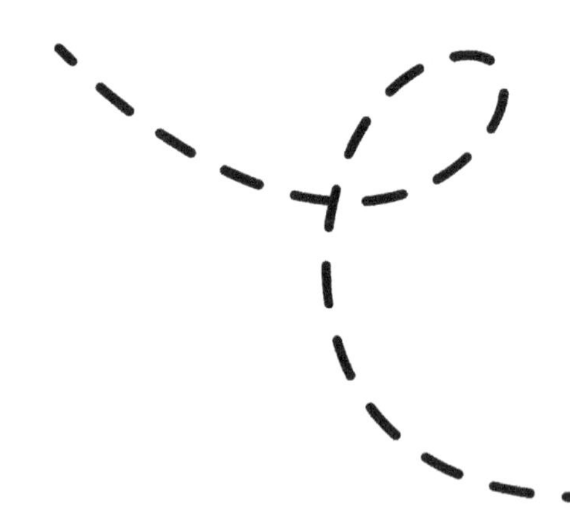

One of the most common questions from schools is this: How can I get more engagement from our followers? The goal is to reach thousands of people each week with your posts, but it is a noisy world out there! How can your school get more of that attention?

Two words: more cowbell!

Oops. I mean these two words: more video!

If Will Ferrell wants more cowbell, then Andrea Gribble wants to see more video on your school pages! Social media favors video. Options such as Facebook Live, Instagram Reels, and Facebook Reels are perfect examples of social media platforms that push video in front of more fans.

Four Tips for Video Content Strategy

If you're thinking about adding video to your weekly content strategy, these ideas will get you started. Sharing a video each week will be easy to schedule [2]

1. Don't worry about perfection. Let me remove some of the stress that may have been caused by my suggestion to add more videos. Your video does NOT have to be a professionally edited video with music, B-roll, fade effects, and graphics. Authentic video from a classroom or event is just fine!

Your video does NOT have to be a professionally edited video with music, B-roll, fade effects, and graphics. Authentic video is just fine!

Now, there is a time and a place for more polished videos, such as enrollment marketing or referendum/bond campaign videos. You can do these videos in-house or hire outside help. But for regular posts, you don't need to worry about a perfectly edited video to share on your daily social media.

2. The shorter, the better! Videos that are seven to ten seconds long actually perform much better than videos longer than one minute. The stats don't lie. Most video views fall off after thirty seconds, many after just ten. So, share that little snippet of students swinging at recess or a child saying they love a book because it is about football and they love the Packers.

3. Share at least one video per week. Upload videos directly onto each social media platform. It's fine if you have a YouTube or Vimeo account for your school, but native videos perform much better. So, on Facebook, upload the video file right onto your Facebook page. Instagram – same thing.

4. Batch some video shoots. Schedule thirty minutes to visit a classroom in your school. Explain that everyone in the class is going to take turns to be on a video for the school's social media pages. Tell the students the question you plan to ask them so they can be a bit prepared with an answer.

Questions may be along the lines of:

- ✔ What do you like most about going to school here?
- ✔ What is your favorite memory about second grade?
- ✔ What do you want to be when you grow up?

Put students in front of the camera, start recording, and then ask the question. Note – if you are interviewing multiple students, make sure that they don't listen to what the other students say on camera. Kids are likely to copy one another, and you'll have ten videos in a row saying that the thing they love most about their school is their teacher. (Although – maybe it's true.)

Time-Saving Video Practices

We've established that video is hot and that reaching your community with real students and teachers telling their own stories can leave an impact much greater than words or images alone.

But who has the time to come up with ideas, shoot videos, then edit and post them? I know it seems overwhelming, but it's easier than you think.

Taking video with your smartphone is perfectly fine and makes it easy to upload onto social media. Set up your camera with great lighting and frame your video with a good angle.

If you'd like some starter ideas to get your creativity flowing, be sure to **go back to the list of one hundred storytelling ideas in chapter 10.**

Get Your Steven Spielberg On

Most people working in schools have little experience in professionally shooting and editing video. However, they are being called on more often to create video content for websites, social media, and newsletters.

Why? Because video works for telling stories!

I reached out to friend and business owner Jake Sturgis, APR. He is the owner of Captivate Media and Consulting, located in Golden Valley, Minnesota. Jake has over twenty years of direct experience working for public schools in a communications role. His team specializes in telling stories for schools. He has a lot of knowledge to share. These are five fabulous video tips Jake has shared to help you master video technique.[3]

1. Have a plan. Hollywood directors don't let the cameras start rolling without a plan. Neither should you. What are you trying to communicate? Who is the audience? How are you planning to distribute the video? If Facebook or Instagram is the main distribution channel, be sure you can incorporate graphics to communicate to the 85 percent of people who watch videos without the sound on.

2. Keep it steady. Grab a tripod or make the world your tripod by finding a table, chair, or a sleeping cat to help balance your camera. Seriously, you may think you can hold the camera pretty steady, but it's tough, especially if you are trying to record for more than thirty seconds at a time. There are light, portable, and inexpensive tripods that will help you take your videos to the next level and allow your viewers to not experience motion sickness.

3. Forget the zoom. Most phones and tablets can capture good video. However, the zoom feature on most of these devices is digital, so when you zoom in, you get pixels! Your video will look extra grainy, and it's harder to keep your camera steady when you're zoomed in on something. Jake's advice is to zoom like your grandpa did back in the day – with your feet! Move closer to the action, and you'll get better-quality shots and better-sounding audio.

4. Record great audio. You can record the best video in the world, but if you are outside and there is a nest of birds chirping right next to the person you're interviewing, your message will get lost. There isn't a good way to easily edit background noise out of an interview. Invest in a microphone or make sure you are really close to the sound you want to record. Take a listen before you start recording. If you can hear it with your ears, the microphone will pick up the same noise. Bad audio equals bad video.

5. Limit special effects. Just because your editing program has one hundred different video effects doesn't mean you need to use them all. Keep the focus on your message. Don't distract your audience with cheesy effects. A standard cut or cross dissolve between two pieces of video should be used most frequently. If you use an effect, there should be a reason, like a "clock wipe" being used to show the passage of time.

The full list of Jake's tips is in a recorded webinar that #SocialSchool4EDU members have access to. The webinar includes elements of a great video story, how to compose shots, how to record high-quality audio, how to edit quickly and professionally, and more.

NINE IDEAS FOR LEADERSHIP VIDEOS

Are your school leaders open to being on camera? Leaders often look for creative ways to make announcements, including snow closings in some states. But in addition to creative school closing announcements, why not make other fun and informative videos?

- Send a customized thank-you message to students.
- Provide an update on a building project.
- Share communication with your staff (note – it's super quick, not a five-minute or longer video).
- Plan regular video updates, like ninety-second interviews with students and staff. You could also record "Talking Tuesdays,"

where your superintendent takes the time to explain responses to frequently asked questions.

- Explain what happens during staff professional development time.
- Deliver a recognition to a staff member.
- Get the community involved in a contest for naming a new school.
- Use Facebook Live for a Q&A before a ballot question.
- Highlight leaders participating in a district-wide event, like a read-aloud day.

You'll find examples of each of these videos from districts across the country in a link in the free bonus resources at socialschool4edu.com/book.

Don't Be Afraid of Live Video

Scared to go live? Facebook Live can be intimidating, but I challenge you to just do it. Once you do it a few times, it will feel more comfortable.

I interviewed Jessica James, a library media specialist from W.S. Neal High School in Alabama, for a *Mastering Social Media for Schools* podcast episode. She shared her best tips and lessons learned for live video. These guidelines will have you streaming live on Facebook like a pro in no time.[4]

You don't have to spend a lot of money to get started. Some simple equipment will up your quality for a low cost.

Jessica uses a **microphone** that costs under $100 and plugs into her smartphone via a lightning cable. This is a shotgun-style microphone that picks up sound from the direction of the source to focus on exactly what you're filming. It also has a furry windscreen to filter out wind noise. Jessica's is a Rode brand. The microphone is powered by your mobile device. Jessica said, "I just keep the little microphone in my backpack that I bring to school every day. It doesn't weigh anything."

A **tripod** is a great low-cost investment to stabilize the phone for a live video. When you're filming live, the way you orient the phone when you go live is how the video will be. You can't change it mid-livestream. So, putting the device on a tripod will also help you think through that step ahead of time.

Letting people know ahead of time that you're going to go live or scheduling a weekly live that they know is coming helps build audience anticipation. Jessica recommended checking out the Wi-Fi ahead of time and making sure that there's a strong signal.

There is no perfect length for a live video. You'll see some benefits in your reach numbers when you're consistent and the community expects a weekly video.

Whether you're showcasing an experiment happening in the third-grade classroom, or some poetry being read in the high school English classroom, or maybe you're at a pep rally or an athletic event, I want you to try Facebook Live. The only way to learn is to try it. You're not going to get any better by being scared of it. Just do it![5]

You'll find a link to the entire podcast episode with Jessica James in the free bonus resources if you'd like to hear more.

HOSTING A DISTRICT FACEBOOK LIVE SHOW

In an episode of my *Mastering Social Media for Schools* podcast, **MORGAN DELACK**, vice president of communications at Finalsite, shared secrets behind the Facebook Live show she did in her former district – how she got started, what tools she used, and the biggest lessons learned. For the weekly show, Morgan tried an interview format first but then pivoted to showcasing what was happening at school: something fun the gym class was doing or a teacher giving a follow-along art demo where a supply list went out in advance.

Morgan was real about the struggles and wins. Ultimately, she encouraged schools to take those risks and try new things for the sake of engagement. She said:

> You have to be OK with failure, too. And I think people understand that you are not a professional television producer, and you are doing this just to give them some fun stuff to look at, really.

> I failed so many times. I had lots of technical issues that popped up where I was talking, but no one could hear me, and I didn't even know that no one could hear me. So, you know, I'd go on and on, and someone would have to run into my office and be like, "There's a technical problem."[6]

Something always went wrong, Morgan said, "but in the end, the feedback was always positive, no matter what happened, so that it encouraged me to keep going and keep trying to improve."

There's no right or wrong way to do it. Morgan said, "I think the communities just appreciate that you're putting yourself out there and you're giving them something that is entertaining and provides value. So, whether you do that like a pro or not, it doesn't matter."[7]

Morgan encouraged thinking outside the box, pivoting as needed, and being as simple or complicated as your skills, budget, and time can handle.

Guide to Facebook Live

The best way to feel comfortable with live video is to have a plan. This doesn't mean over-scripting. It means ensuring everyone who may appear in the video knows it will be live. It includes selecting content that will translate well to video.

One week before:

- ❏ Practice on your personal page first.
- ❏ Get access to post on the Facebook page.
- ❏ Check the Wi-Fi signal in the location that you plan to broadcast from.
- ❏ Identify children who can't be online.
- ❏ Promote it so people know the date and time (optional – spur of the moment is OK, too).
- ❏ Locate the tripod or monopod so you can minimize shaky video.

- ❑ Consider free prizes for viewers who comment. This works well for Q&A sessions.
- ❑ Don't worry about scripting the entire broadcast.

Minutes before:

- ❑ Make sure your phone or device is fully charged.
- ❑ Last-minute potty break! (You'll thank us later. Promise.)
- ❑ Ten minutes before, get on the Facebook page, scroll down to where it says "Go Live," hit that button (yikes – don't push the record button yet!), make sure everything's working OK, get in your position, check the lighting, and make sure you can hear well. Then, exit and calm those nerves until showtime!
- ❑ About two minutes before LIVE, hit "Go Live," get in your position, choose if you want to record your video in vertical OR horizontal layout (in selfie mode, mirror the camera so wording on shirts or background can be read), enter a descriptive title, then push the red record button.

During recording:

- ❑ Smile, have FUN, and be real. People love authenticity!
- ❑ Encourage people to comment.
- ❑ Thank viewers for watching.
- ❑ When you're finished, choose "Post" it if you want it available for future viewers. Posting publishes the video to your timeline so if people didn't see it live, they can still see it in their feed later on. With certain items, like concerts or book readings, you may not be able to post it based on copyright or fair use law.

Grab It!

Get printable PDFs of the lists above in the free bonus resources at socialschool4edu.com/book. Look for **"Guide to Facebook Live"** and **"30 Ideas for Facebook Live."**

30 Ideas for Facebook Live

Sometimes it helps to have some ideas to get you started. Here are thirty that nearly every school can do. After you try out Live on the obvious ones, try some other options and watch your engagement increase.

Obvious reasons to broadcast

1. Concert (may need to check music copyright rules with your music department)
2. Sporting event (see more below)
3. School board meeting
4. School play (check copyright rules)
5. Graduation
6. Special speaker

Features at sporting events

7. Pep band or special music before the game
8. National anthem, especially if students are singing
9. Game breaks: special recognitions, dance, or performance
10. Student section having fun
11. Player or coach interview after the game
12. Big game? Film students saying "Go Eagles" or "good luck"

During the school day

13. Kids in the hallway, headed to a special event
14. Front entrance, especially first or last day of school
15. Lunchroom "go live," especially if it's a special food day
16. Classroom experiment
17. Art class
18. Gym class, film a fun or unique activity
19. Music or drama class, promote an upcoming event
20. Academic classes: speeches, projects, games, and more
21. High school classes, pan through projects in family and consumer science, tech ed, or computer classes

Special events

22. Field trips
23. Staff recognition

24. Referendum town hall meetings, Q&A sessions
25. Tours of remodels or additions to your school

Nighttime activities

26. Promote non-sport activities: drama, after-school programs, quiz bowl, and more
27. A teacher/principal goes live reading a bedtime story (get permission from author/publisher first)

Announcements

28. Broadcast school donations
29. New logo or website reveal
30. Weekly or monthly update from superintendent or principal

YOU'VE GOT THIS!

- [ ] Start doing more video – once a week at least.
- [ ] Stop worrying about being perfect.
- [ ] Don't worry about special effects.
- [ ] Keep pre-recorded videos short for the best engagement.
- [ ] Make use of budget-friendly equipment for better audio and quality.
- [ ] Take the plunge into trying live video!
- [ ] Consider a weekly Facebook Live or a weekly show.
- [ ] Focus on celebrating students and staff and telling great stories!

COLLECTING AND REPORTING METRICS

17

Please don't scroll by! Metrics doesn't have to mean misery. I'll show you an easy way to create a report card that shows your leadership team the power of social media!

Case Study

Meet Jenny Starck: Jenny is the superintendent of Cadott School District in Wisconsin, a district of around 850 students in K–12. Social media is important for her district and community because it's a positive space to celebrate their schools.

The challenge: Only admins and the account manager from the #SocialSchool4EDU team can see the numbers on social media metrics. But the school is reaching the community in big ways! Sometimes the reach goes beyond expectations.

The process: Every quarter, Jenny gets a "report card" from the #SocialSchool4EDU team that gives a snapshot of where Cadott School District is with social media metrics and where it falls in line with other schools. It shows the posts with the most reach, too. Jenny shares that report with staff at a recurring meeting.

The outcome: The school uses some of the data in its communication plan, and board members receive data annually, if not more often. Showing staff how the district trends in comparison with other schools of a similar size has sparked discussion, especially when they see a post with a reach of 7000 people in a town that only has 1300. It also inspires a little friendly competition with other schools in wanting to stay "ahead." Competition sometimes leads to staff submitting more content for posts.[1]

NEW AUBURN

SOCIAL MEDIA REPORT CARD » AUG - OCT 2022

facebook

Total Followers: 2,609

Growth: +264

Average Monthly Reach: 12,800

New Auburn Facebook - Monthly Reach

Instagram

Total Followers: 885

Growth: +36

Average Monthly Reach: 4,967

twitter

Total Followers: 208

Growth: +0

Average Monthly Tweet Impressions: 846

WE NEED TO
MAKE THE POSITIVE
SO LOUD
THAT THE NEGATIVE BECOMES
ALMOST IMPOSSIBLE
TO HEAR.

POWERED BY #SOCIALSCHOOL4EDU | WWW.SOCIALSCHOOL4EDU.COM

Social media makes a difference. I've seen it work in positive ways in hundreds of schools and districts! But if you're not reporting on that impact to your leadership team, your board, and your staff, they may not realize the power that is possible.

In this chapter, I'm sharing how easy it is to create a one-page social media report card that you can produce in less than an hour. Really!

Gather Your Stats

The amount of analytics available on social media platforms can be overwhelming. Through trial and error, I've zeroed in on the stats that are the most important. Of course, you could choose to share more, and you can report on more platforms than just Facebook, Instagram, and Twitter.[2]

Step One: Set up a simple spreadsheet so that you can record these numbers each month. Our team uses Google Sheets.

Step Two: Set up a monthly reminder to grab the metrics you need. **You can plan to look on the first day of the month and grab the previous month's data.**

Facebook: Find these numbers within "Insights" by viewing your Facebook page on a computer, or on your phone using the Business Suite app. From there, you can record the following categories for your report:

- **Total followers** – Facebook makes it confusing. It lists both "likes" and "followers." With the latest update, Meta indicated that "likes" would be going away, but so far, I can still see it. The only difference is that someone can follow your page without liking it. That is why followers is usually a bigger number. Whatever you choose to use, just be consistent!

- **Growth** – The number of new followers you've added since the last report card. Once you start tracking your data every month, this number can be calculated through simple subtraction.

- **Monthly reach** – Facebook provides a monthly reach.

- **Best Posts** – Scroll through your posts and look for the two posts that reached the most people. The Business Suite app allows you to sort by reach, as well.
  - » Save the best image from each of the top two posts.
  - » Record the dates and how many people each post reached.
  - » Briefly summarize the posts.

Instagram: Make sure that your Instagram account is set up as a business account and that it is connected to your Facebook page. This gives you access to insights on both the Instagram app and the Business Suite app. From there, you can record the following categories for your report:

- **Total followers** – The number of people who follow your Instagram account.

- **Growth** – The number of new fans you've added since the last report card. Once you start tracking your data every month, this number can be calculated through simple subtraction.

- **Total Reach** – Instagram provides a monthly reach.

Twitter: Currently, Twitter analytics are found under Creator Studio. You can view the entire previous month. Here's what you can record:

- Total followers

- Growth

- Tweet impressions

DATA SPEAKS

AMANDA KELLER, social media director for Weston School District in Wisconsin, uses metrics to show other school leaders which types of posts get the most engagement. Whether they're building web content or posting on social media, she can show what works. Amanda's message to staff is, "I want to make you look as good as possible because that's my job, and then we all are winning by telling our story while I make you look like this amazing teacher in this amazing school – which we are."[3]

Create Your Report

Now that you have the numbers, it's time to create a report. As you do the following steps, keep these things in mind. First, make the report visual and fun! Including photos from top posts will help draw attention to the information you're sharing.

Second, keep it short. You could easily fill up a few pages or a whole slideshow with metrics and photos. But the chances of people reviewing pages of data are not high. When I was just starting in social media management for schools, that's what I did. Every month, I had a bunch of slides that I wanted to review with the leaders, but they just didn't have time. Keep your report to about one page to pack a bigger punch!

Okay. I promised you could do this in under an hour. So here are the steps.

Make the report visual and fun! Including photos from top posts will help draw attention to the information you're sharing.

Step One: Choose your favorite design program to assemble your visual report. Canva is my favorite tool, and it was used to create the report example shown here!

Whichever tool you use, create a template you can simply fill in each month. Here's a time-saving note: those who are part of my membership group have access to a Canva template they can customize with their school branding!

Step Two: Fill in your report template with the data you've gathered.

Step Three: If you have room at the bottom of your report, you could also add a little "ask" for your staff. Such as: "Remember, submit your photos and stories to be featured by emailing them to socialmedia@ourschool.org. In February, we are focusing on what we love about working in this school district. In two sentences or less, share your story!"

See a sample report on page 220.

Share Your Report!

Ideally, you should spend a few minutes sharing the social media report at a school board, cabinet, or staff meeting. If you can't get on the agenda, at least email the report to all staff and board members.

You could even consider sharing your report on your website or in a newsletter to families. It's a great reminder to follow the social media accounts!

Be consistent. This report isn't something you can do just once in a while. I'm serious when I say this entire report takes less than an hour to create. So, commit now to doing it every month this year. That's just twelve hours of your time over the next twelve months – and the time it can end up saving you is ten times this investment.

How does it save time? Well, if this report leads to staff submitting more content, you don't have to spend hours each week trying to take photos and come up with your own stories. This will get you hours back every single week!

Use Metrics to Plan Future Content

When you use social media insights and analytics to create a monthly report card, you can also use those metrics to create future content that will hit the bullseye for your audience. Examples of how that information can give you content ideas:

- ✔ Facebook posts with the most reach
- ✔ Facebook posts with the most engagement
- ✔ Facebook post types
- ✔ Top Tweets
- ✔ Top Twitter follower
- ✔ Most Instagram reactions or views

When you see what gets the most engagement, it tells you to share more of that type of content.

Grab It!

How do you hit the social media bullseye using analytics? Find out in this short YouTube video that gives you a visual on how I collect insights and use them to drive future content. Grab the link to **"Hit the Social Media BULLSEYE Using Analytics"** in the bonus resources.

One blog commenter reacted to the social media report card tutorial on #SocialSchool4EDU and said, "I LOVE THIS!! This is going to be so perfect to present to my administration who is not on social media."[4] Yes! A simple report can help you highlight the importance of your work.

Best of luck as you showcase the power of social media!

YOU'VE GOT THIS!

- ❑ Metrics, analytics, or insights don't need to be intimidating.
- ❑ Track a few numbers in a spreadsheet each month: followers, growth, reach or impressions.
- ❑ Create a template to visually share your metrics each month.
- ❑ Share that graphic with staff, board members, and community.
- ❑ Use the metrics to plan future content that hits the mark.

Would your colleagues find this book helpful?

Share a link to socialschool4edu.com/book

PROFESSIONAL DEVELOPMENT

AVOIDING BURNOUT

18

If you try to be all things to all people, that's going to take a toll on your personal wellness! If you commit to intentional practices to avoid burnout, you'll be a better leader and enjoy less stress, too.

Case Study

Meet Christy McGee, APR: Christy is the director of communications for Fountain-Fort Carson School District 8 in the Colorado Springs area. The district has fourteen schools with around 8000 students. She's been in communications for seventeen years, serving the last seven of those in FFC8. She has degrees in communications and sociology. Christy was a one-person shop for a lot of years until the district recently hired a strategic communications specialist.

The challenge: Christy said something about the COVID pandemic woke her up to the importance of mental health, wellness, and work-life harmony. In the summer of 2020, she found herself needing to set boundaries and break a pattern of working 24/7 for more than eight weeks. One part of that motivation came from her kids' comments about being "on your phone all the time." She said kids are "a pretty good meter" to indicate if you're "working too much at home."

The process: Listening to podcasts, reading K12prWell newsletters,[1] and attending sessions at conferences, Christy learned some tips for finding the balance she needed. She turned off work notifications on her personal phone. She said one of the "biggest game changers" for working from home was not having her phone pinging to check it all the time. She still gets work emails on her phone and uses the social media apps, but she can check it on her time schedule instead of having notifications all

CASE STUDY

day – including during off time. She also set boundaries for her time.

The outcome: "I have a boss who also talks this talk and walks this walk," Christy said. He reminds staff not to check email while they have a vacation and gives staff freedom to say when they have other commitments, making it comfortable to put a weekend request off until Monday. Christy said, "It feels scary. It feels like we need to be available, but if you can manage the expectations before a crisis comes up, then it's a little less scary. It's about managing other people's expectations of you."

Christy was able to take a two-week trip to Africa without checking anything work-related. She planned ahead, making sure her team had connections for who to contact if questions came up. Sometimes that's a contact in another district who can help with questions specific to software or platforms. She has an emergency sub plan in place with all the details needed. "To avoid burnout, it's just really prioritizing yourself. . . . It's not just at home; it's also at work," Christy said. She now enjoys being productive at work in a way that also fills her bucket and recharges her battery.[2]

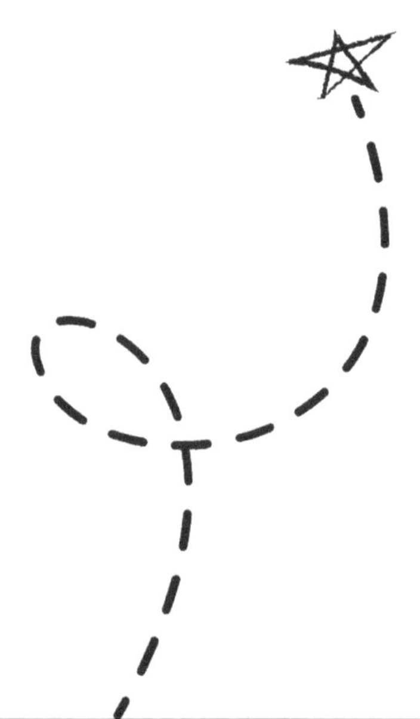

As a school communicator, you fill an important role. This role was deemed *critical* when schools closed their buildings at the start of the 2020 pandemic, and it's a role that continues to show its importance as we navigate through this school year.

But with the never-ending requirement for constant communication in a 24/7 news cycle, stress and overwhelm are a daily reality. Still, we want to manage social media, not have *it* manage us!

So how do you build a rewarding career in school communications without sacrificing your personal life?

I don't have all the answers, so I reached out to school communicators across North America for their best advice. You'll see their advice boiled down into ten simple tips below. Plus, they shared some advice in their own words.[3]

10 Practical Tips to Avoid Burnout

For things to get better, you have to be willing to commit to at least a few of these ideas. My promise is that if you take action, you will experience a better work/life harmony. I don't think "work/life balance" is the right word, but work/life harmony is absolutely possible. It can get better.

1. Set boundaries for your workday. Only respond to emails and phone calls during your normal working hours. Yes, emergencies will happen. Your leadership team should be aware of how to reach you if something is truly an emergency, but that method should not be shared with everyone in your district.

There will always be "one more thing" on your to-do list. To help you shut things off when you're not working, create a list of priorities for the next workday and leave that list at the office.

> Acknowledge that there will always be items on your to-do list at the end of the day. Of course, you could stay an extra hour or two or work from home a few hours after dinner, but guess what? Tomorrow morning there will still be a to-do list. Prioritize what must get done and go home at the end of the day! Set the boundary and be present when you are away from work. . . . Remember, you are not your job or your career or your position. You're a human. You deserve time away each day to be with family, friends or just to be you.[4]
>
> —**MARCIA KELLEY**, School Information Officer, East Syracuse Minoa CSD, New York

2. Schedule communications in advance. Most social media posts do not need to be posted the moment the photo is taken. Use the scheduling features on social media to your advantage. You will reach more people if you utilize the four best posting times and limit (or eliminate) weekend posts.

Waiting on the results of a big sporting event? Get your AD and coaches involved by asking them to use the school hashtag on Twitter so you don't have to be responsible for tweeting every update in the moment.

Your email communications can also be scheduled in advance, so take advantage of that feature in your software.

3. Talk it out. Your role comes with a lot of stress and pressure. Find a trusted person to talk to, such as a trained counselor, your spouse, a good friend, or another school PR colleague. They don't need to have all the answers to your struggles, but verbalizing what is on your heart will help.

4. Take breaks in your day. Fresh air and exercise should always be a priority. It's amazing what a ten-minute walk can do to spark your creativity for your latest project!

"I had to learn to set boundaries, almost like baby-proofing a room, to protect my mental health. I put bumpers on my day . . . I start the day with a walk, exercising in quiet as the sun rises, then sit on my living room floor doing a devotional, prayer, and meditation (sounds loftier than it is . . . a lot of it's just sitting quietly and relaxing my mind and body). Only after that, after I've gotten my mind in the right place, do I dare look at email and social media. At the end of the workday, I prioritize time with my family, doing things that bring me joy and relaxation – and unless it's an emergency, I don't check social media or email at least a good hour or two before bedtime. It's a conscious, constant effort to prevent the work from creeping into my private life. I will confess that melatonin has become my friend in getting a good night's sleep."

—**MICKEY SCHOONOVER**, APR, Director of School-Community Relations, Pattonville School District, Missouri

5. Choose which requests are urgent. Establishing deadlines is critical to success in your role. It's easy for you, as an overachiever, to assume everything is needed ASAP. But define clear deadlines with your colleagues so they have realistic expectations for your work. They may not even realize all the responsibilities on your plate, so have an honest conversation about what you're working on and how their request fits into your priorities.

"The best tip I ever got was not to assume that every request you receive is urgent. It's so easy to FEEL like people expect you to deliver everything right away, but I am often surprised by the results when I say something like, 'I'm finishing up another project right now. Can I get that to you on XXX date?'"

—**EMILY F. POPEK**, Public Information Specialist, Capital Region BOCES, New York

6. Have a backup. The myriad of things you are responsible for in your district should each have a trained backup person. Taking time off is important, and you also need to be replaceable in case of an emergency. Posting on your school's social media while you are on vacation is not acceptable.

You do not need to post on social media pages while you are on a personal vacation. Let me say that louder for the people in the back – you do NOT need to be the one responsible for posting when you are on vacation. Assign someone else to post while you are away. If you have the email system set up that I taught you earlier in the book to receive submissions, this will be easy. Someone else can check that email while you are away.

7. Don't take things personally. Because you're responsible for so much of the communication in your district, it's very easy to take things personally when people have questions or choose to criticize the district. Remember that you are just the messenger. Do not take these comments and complaints as personal attacks.

8. Practice empathy. Understanding will get you everywhere. You work in school communications because you have empathy. Showcase that special trait in each interaction that you have – with staff, with students, with parents, with board members, and with the community.

Habit 5 in Stephen Covey's *The 7 Habits of Highly Effective People* is "Seek First to Understand, Then to Be Understood."[5] This will get you far!

9. Remember that your words matter. When giving recognition in your public communications, it's easy to just focus on your rockstar teaching staff. But remember that the words you use matter. There are just as many support staff and administrative positions that help your school achieve its mission. Make a conscious effort to use words that praise your entire staff.

And this should go without saying: do not ever bad-mouth others.

This includes staff members, parents, community members, and members of the board. Everything comes back around. Hold yourself to a high professional standard.

10. Put away your technology at home. The physical act of putting your phone in a drawer or in another room will help you focus on your family and personal life. If you must have your phone or laptop with you in the evenings, let your family know you're working. It's easy for kids to complain that "Mom is on her phone again," but if you explain that it's part of your job and let them know that tonight you have to help respond to some questions, they'll understand.

Still, refer back to number one! You don't need to be "on" all the time.

I know it isn't easy to change ingrained habits, but you absolutely need to have boundaries and standards in place. Your health and sanity need to be a priority.[6]

Grab It!

You can download a shareable PDF with these ten tips from the free bonus resources at socialschool4edu.com/book. Look for **"10 Practical Tips to Avoid Burnout in #SchoolPR."** This would be a great list to print out and place it in your office as a loving reminder to take care of yourself!

Prioritizing Self-Care

In a *Mastering Social Media for Schools* podcast conversation I had with Kristin Magette, APR, she talked about how her experience with trauma had a dramatic impact on her career. She also talked about how empathetic leaders experience challenges when they deal with traumatic things at work. When students, families, and staff deal with trauma, it affects us.

Kristin said, "We absorb a lot of that shock so that we can communicate with our community – whatever part of our community – and not traumatize them. We're trying to minimize the trauma." She compared it to the shock absorbers on a car that has traveled a lot of bumpy roads. Those shock absorbers don't last forever, she said.[7]

Boundaries are at the top of the list for protecting your work/life harmony.

We need to be aware that being good at this aspect of our job comes with rewards and consequences. Kristin shared some strategies for dealing with this stress and protecting ourselves:

- ✔ Boundaries are at the top of the list for protecting your work/life harmony. They happen through conversations with coworkers, not automatically. Know what you need to protect so you don't have a breakdown. Examples:
  - » Turn on the "do not disturb" on the phone in the evening and overnight. Having a conversation ahead of time lets your supervisor know when you can't be reached or how to reach you with something truly urgent.
  - » Turn notifications off on your phone to reduce pings, buzzes, and beeps.
  - » Recognize negativity and which conversations or groups you need to personally step away from.
- ✔ Work with a counselor.
- ✔ Slow down your physical body. Notice how fast you're doing everyday tasks – even brushing your teeth. Kristin referred to it as the comical motion of brushing teeth "like an ostrich trying to fly." She practices walking in slow motion sometimes. She introduces what she calls periods of "landing" between periods of "flapping." This means she intentionally slows down or stays still for a period of time between the busy parts of her day that she must move at a faster pace.
- ✔ Find activities that recharge you and make them regular habits.

Kristin said we need to understand that "if we care for ourselves and our family, we can actually serve our communities better. . . . We can actually serve our students and our teachers and our staff better." She said there is no substitution for caring for yourself. Ultimately, it's what makes us the best leaders in our jobs.

EMERGENCY SUB PLAN

I've mentioned the importance of having an emergency plan for having someone else cover your role if you need to step away. Kristin Magette, APR, shared some of her story in the podcast interview above, and even more in a blog post for #SocialSchool4EDU. (You'll find Kristin's full story linked in the bonus resources.) The bottom line is Kristin wasn't prepared to spend time with her son in the pediatric ICU at the height of the back-to-school season. Her experience revealed the lack of an emergency sub plan, which prompted immediate action when she returned to the office. Here's what Kristin determined should be included in that plan.

- ✔ An annual calendar of her recurring, predictable events, programs, deadlines, and other responsibilities.
- ✔ A list of her logins or cross-trained employees for technology systems, including social media accounts.
- ✔ A list of regular outside contacts – vendors, customer service, and in her case, the names and cellphone numbers of a couple of trusted PR professionals.
- ✔ A commitment to keep her work calendar (Google, Outlook, etc.) updated every day with meetings, deadlines, and other projects.[8]

Take the time and prepare your main responsibilities, share your social media passwords, and keep that calendar updated. Sometimes, life takes you out of circulation for a bit, and everyone benefits from being prepared for the worst.

Grab It!

Tanisha D. Hearn, communications specialist for Seguin Independent School District in Texas, shared how she manages her time in podcast episode 134 of *Mastering Social Media for Schools*. Look for **"Give the People What They Want with Tanisha D. Hearn"** in the bonus resources.

You Don't Have to Do It All

Perhaps you need me to say it outright. You can stop posting on weekends and thinking everything has to be posted immediately.

I've been managing social media for schools since 2014. I've learned a lot in that time frame – most importantly, that you cannot be online 24/7. If you feel as if you need to post everything immediately and that you need to post on both Saturday and Sunday, I want to give you permission to give yourself and your fans a break!

We see remarkably lower engagement on our social media channels on the weekends. Why? I'm not entirely sure, but I'm assuming it's because our fans are out living their lives! They are focused on family life and doing activities away from school and work, and they aren't as active on their phones.

So, if you're still posting on the weekends, I want to challenge you to stop. Now, if you have a big event or playoff game happening, you may make an exception. But the majority of your posts can truly be scheduled for Monday through Friday. Your engagement will probably increase!

The majority of your posts can truly be scheduled for Monday through Friday. Your engagement will probably increase!

YOU'VE GOT THIS!

- [ ] Be intentional about protecting your time, energy, and attention. Set boundaries for availability and discuss them with your team.
- [ ] Take breaks to move and stretch. Take a walk!
- [ ] Have an emergency sub plan.
- [ ] Practice empathy and understanding with others while being aware of how absorbing "shock" also affects you.
- [ ] Remember that everything isn't urgent.
- [ ] Slow down. Recharge.
- [ ] Reach out to others for help and encouragement.

YOUR PROFESSIONAL LEARNING COMMUNITY

19

Social media is always changing. You don't need a degree to serve as a school communicator, but don't skip the professional development and networking opportunities! Not sure how to fund it? I'll give you practical steps to make it happen.

Case Study

Meet Megan Anthony: Megan is the communications coordinator for Canal Winchester Schools in Ohio. She previously served as an education reporter until someone urged her to apply for a school administrative assistant and school communicator combination job. Since then, she's been in several professional roles for school communications.

The challenge: "Being a communicator at a school district can be a lonely job," Megan said. Most districts have one person in the communications role, and others on staff don't care all that much about social media. When the loneliness was starting to get to Megan, she looked for support.

The process: Megan started listening to podcasts related to marketing, communication, and school PR and discovered a podcast that resonated with her. She learned about #SocialSchool4EDU on a podcast and joined the **membership program.**

The outcome: The camaraderie Megan found in connecting with other school communicators made her excited to go to work again. "It was a turning point for me," she said. Now she's part of the #SocialSchool4EDU membership program, which didn't exist yet when Megan first reached out to connect with others. She's part of NSPRA, a #K12prchat, and another pro Facebook group. Networking is key, she said, because "everybody becomes better by talking to people who have different perspectives and ideas – and learning from

CASE STUDY

people who are further along than them, who are newer than them, who are at the same level as they are right now."

Being part of associations helps Megan with her professional development and has also given her personal growth. "I've just gained a lot of confidence in addition to all the wisdom and resources and support I have seen, and I have gained so much confidence in my abilities as a communicator. And I really do know things. I'm not just making it up all as I go, as much as I feel like it some days," she said. "Everybody needs a friend sometimes."[1]

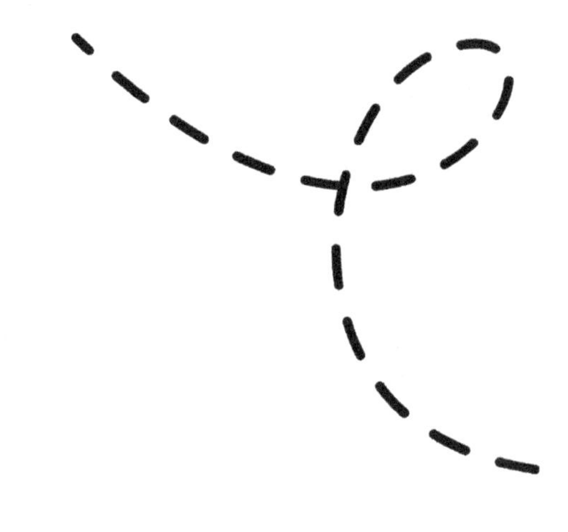

SCAN QR CODE

FOR BONUS RESOURCES

socialschool4edu.com/book

After analyzing thousands of school social media pages, I've come to identify one key factor that will determine the success of social media presence for schools. Hands down, it's having a person dedicated to staying on top of the game – to continually learn and consistently push for out-of-the-box thinking when it comes to telling your school's story.

After all, social media is not something that you just learn and then you're good to go forever. It's constantly changing – sometimes daily! If you don't continue to keep up with those changes, your school will be left behind regarding the true engagement that you can have with social media.

The Always-Changing World of Social Media

What do you struggle with the most? Can you nod your head to any of these?

- Struggling to find time to manage social media on top of all the other things you do in your role
- Getting staff to let you know about the events they want to be featured on your social media channels
- Feeling frustrated with Facebook constantly changing its layout
- Monitoring the comments and worrying about negative feedback
- Dealing with complaints that you're not "fair" in the way you cover certain students/staff/activities
- Spending way too much time trying to develop great graphics
- Trying to figure out all the ways you can share content on Instagram

- Wondering how to create a quick video that combines all the short videos and photos you just received from several staff members

Social media is not something that you just learn and then you're good to go forever. It's always changing – sometimes daily!

You're not alone. Questions like these are why I started the *Mastering Social Media for Schools* podcast. It's also why we have a huge library of content and tools on the #SocialSchool4EDU blog. If you have a school social media question, there's probably an answer in one of those blogs.

I know you are BUSY! You have so much coming at you during the day that you don't have time to read one more blog or watch one more video or follow one more Twitter account. That was part of my motivation for starting a podcast. Podcasts are different because you have time to listen while you're driving, while you're working out, while you're folding laundry, or while you're prepping dinner.

If you can turn your car into a classroom, your workouts into some career inspiration, or your chores into productive learning time, you'll actually feel like you're getting ahead instead of falling behind!

Your time is precious and limited, but let's look at some other ways you can stay on your game.

Tips for Staying Current

- Connect with other school communications professionals.
- Join the #K12PRchat that happens on Twitter.
- Listen to podcasts.
- Read blogs (thanks for reading the #SocialSchool4EDU ones!).

- Watch other schools' Facebook, Twitter, and Instagram pages.
- Attend online webinars, many of which are free, run by organizations and businesses.
- Follow businesses and brands to see what they are doing, and then apply those ideas to your school.
- Go to conferences for school public relations – like NSPRA or your local state chapters such as MinnSPRA in Minnesota, WSPRA in Wisconsin, or OKSPRA in Oklahoma.

Upping Your Game

Continuous learning is a common concern for just about every school communicator I know! Why? Well, you work in a school. Professional development (PD) is pretty much a given for all educators and education professionals. You're surrounded by colleagues who are constantly attending PD events, earning additional credentials, and upping their game when it comes to new approaches in education.

On top of that, you work in a demanding field. Communications best practices change quickly, not to mention all the technologies you're asked to juggle on a daily basis. Between crisis communications, social media, internal communications, and external communications, there's a lot to know!

That's why PD for School PR professionals is absolutely critical. I had the privilege to interview Janet Swiecichowski, EdD, APR, Fellow PRSA, and vice president at CEL, for a podcast episode on *Mastering Social Media for Schools* all about PD for school communicators.[2] Janet has made a lifelong commitment to professional development. She's been in school PR for over twenty years in various capacities and is a fierce advocate for continuous learning.

Getting PD at the national level is important, but it's equally critical to make local connections with like-minded professionals. For example, many media outlets contacting schools are regional, so it's helpful to have colleagues to bounce ideas and questions.

That's why NSPRA also offers regional and local opportunities for PD and networking through state chapters. When I started #SocialSchool4EDU in 2014, I started out with the Wisconsin chapter (WSPRA) and later joined the Minnesota chapter (MinnSPRA). As it turns out, Janet is a past-president of both chapters.

To strengthen her public relations chops, Janet also joined the Public Relations Society of America (PRSA), which offers a vast archive of webinars, courses, and certificate programs.

Janet earned her Accreditation in Public Relations (APR), a certification that requires rigorous coursework. She described it as the "gold standard" in the public relations field.

The APR is worthwhile for any school communicator who wants to move beyond the tactical and into the strategic within their school PR role, according to Janet.

ABOUT NSPRA

Since 1935, the National School Public Relations Association (NSPRA) has been providing school communication training and services to school leaders throughout the United States, Canada, and the U.S. Department of Defense Education Activity (DoDEA) schools worldwide. As the leader in school communication, NSPRA serves more than 2,500 members who work primarily as communication directors in public school districts and education organizations. NSPRA provides high quality professional development programming through on-demand learning, an annual National Seminar, webinars, online forums, and resources.

NSPRA's mission is to develop professionals to communicate strategically, build trust, and foster positive relationships in support of their school communities. That mission is accomplished by developing and providing a variety of diverse products, services, and professional development activities to association members as well as to other education leaders interested in improving their communication efforts.[3]

"Public relations is about building mutually beneficial relationships with the people upon whom our success or failure depends for any organization."

—Janet Swiecichowski, EdD, APR, Fellow PRSA, Vice President at CEL

CARA ADNEY, a ten-year veteran of social media management, was part of a pilot program for a boot camp put on by the #SocialSchool4EDU membership group. In a podcast episode of *Mastering Social Media for Schools,* Cara shared how being part of the boot camp helped her career, even as an experienced manager. Professional development matters! Cara said the five-week program included time to learn, ask questions of the leader, process, ask others in the group questions, and get input from people at all stages of social media strategy development.

Cara said she benefited from a challenge to establish a system. "I've known for ten years that I needed a system," she said. "Once I started planning out my calendar and making a list – like each month, these are the events that our students are in – I honestly told my boss, 'It's no wonder I'm overwhelmed.'" It changed Cara's perspective and helped her figure out where she needed to work ahead and create a plan to streamline. The best feedback was when her boss said to her, "I can tell this is good for your skills, but most importantly, it's been good for your soul." That's the benefit of learning in community, no matter how experienced we are![4]

State and National PD Programs for School PR Professionals

Janet earned her undergraduate degree in communications, but when she started working at a school, she realized she needed to learn more targeted methods. Her superintendent connected her with Rick Kaufman, APR, who encouraged her to join NSPRA.[5]

This membership helped Janet get connected with people, opportunities, and learning. She's even gone on to present at the annual national NSPRA conference for nearly twenty consecutive years!

Higher Education Opportunities for School PR Professionals

Janet Swiecichowski didn't stop at state and national groups when it came to her professional development. She took graduate classes in public relations to augment her general communications background, later earning her master's degree in strategic communication from the University of Minnesota – Twin Cities.[6]

In fact, to this day, she is still investing in her own higher education! At the time of publication of this book, Janet had just completed her Doctor of Education (EdD) in organizational change and leadership.

Yet don't be alarmed. You don't have to commit to a full-fledged higher ed program! You could take classes at your community college or local university that are practical, hands-on, and teach you the latest techniques in digital communications and tactics. For example, you could take a course on video production.

Other free or low-cost programs for communications-related PD include Google Analytics, Hubspot, and Social Media Examiner.

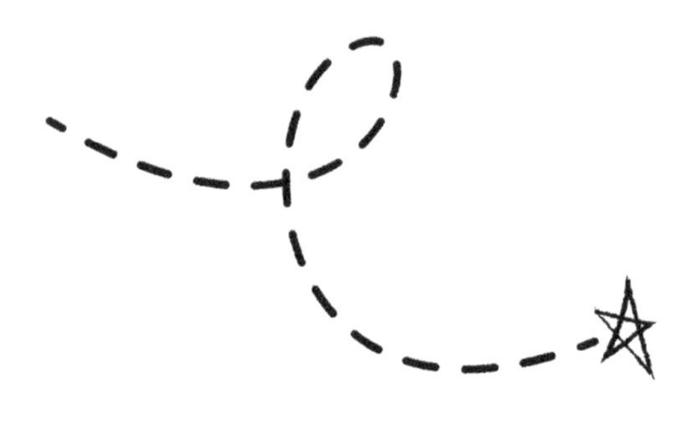

Professional development should be ongoing.

Other PD Opportunities for School PR Professionals

Janet stressed the importance of taking care of yourself and your well-being before pursuing these programs and memberships. You work in a high-stress environment with many pressures coming from all directions.

#K12prWell, a movement co-founded by Kristin Magette, APR, and Shawn McKillop, APR, connects the school PR community through wellness and self-care initiatives for avoiding occupational burnout. Kristin and Shawn also lead webinars to help school PR professionals learn how to invest in their well-being.[7]

There's value in a "one and done" certificate or degree program, but I recommend that you pair it with some sort of ongoing membership or PD program.

At #SocialSchool4EDU, we've developed a program to help school communicators improve their skills – specifically, in social media. Our virtual membership program includes year-round training and support so you will stop being overwhelmed and start enjoying your job. Learn more about it in the next chapter.

Grab It!

Get a list of all state chapters of the National School Public Relations Association at the link in the bonus resources.

MENTORSHIP

Thanks to the availability of worldwide virtual connections, we can find mentors just about anywhere if we're looking for help. Here's an example of what that can look like, based on a conversation I had with Susan Brott, APR, in a podcast episode of *Mastering Social Media for Schools*.

After Susan received the 2021 Barry Gaskins Mentor Legacy Award from NSPRA, she happened to be responding to a Twitter post that NSPRA had shared about the award when a commenter said, "I would love a mentor." That connection developed into a mentoring connection that includes:

- ✔ Monthly check-ins
- ✔ Text messages back and forth
- ✔ Mutual learning for both the mentor and mentee
- ✔ Long-term encouragement and support
- ✔ Accountability
- ✔ Networking opportunities

Susan said she does this because of how mentors have shaped her personally and professionally. This is her way of giving back and paying forward. She's also been known to get off a client call with fresh energy and motivation. "I don't think I'm necessarily imparting all this amazing wisdom," she said. "It's just having the conversation and having that connection with people."[8]

NSPRA currently has a formal mentor program that is free for members.

Grab It!

Check the bonus links for the one-page flyer for our membership program as an example. If you don't spot one of these for the program you're looking at, ask for it. They likely can provide it for you.

You understand the value of PD, networking, and membership, but you're probably asking, "Who's going to pay for it?"

It seems like we never have enough funds in the budget to do everything we need to do for our students. We feel guilty focusing resources on ourselves because we're so used to giving to others. And we also think we have the internet and YouTube. If we don't know something, we can Google it or watch a video!

Your learning is an investment in your school, not just personal growth, so I want to show how you can prove to district leaders why it should be part of the annual budget.

7 Steps to Get Funding for PD Programs

The best, most impactful school communicators out there invest in professional development for themselves – whether it's through state or national organizations, product-specific training like InDesign or Adobe products, or social media training like the programs #SocialSchool4EDU offers.

I've put together seven helpful steps to get professional development funding approved for your role.[9]

Step 1: Identify the Professional Development Program

This seems obvious, but if you are asking for money to sign up for a program, you'd better know what's included, the cost, and the time commitment. There are many opportunities available, so if you're not clear on what you want, it's going to be tough to get approval for it.

Many programs will have a one-page flyer that will detail this information, making it much easier for you to understand and to explain it to someone else.

Next, read the testimonials. A good program will provide feedback from current customers so that you have insight into what kind of results you can expect.

Step 2: Evaluate Your Strategic Plan

How does professional development tie into the strategic plan? I can almost guarantee that "communication" is in there somewhere, so communications-specific PD shouldn't be too tough to tie in.

If you don't have a strategic plan in your school district, that's OK. Do you have overarching goals or initiatives? Are there some common themes that the board is looking to improve? That's what you need to research so that you can tie your chosen program to some high-level goals for your school or district.

Shane Haggerty, APR, former school communicator and current managing director of marketing and communications at Battelle for Kids, made the perfect point: "If you are employed within a school district, your professional knowledge and skillset are part of the strategic plan. What you bring to the table in helping the district achieve the components of its strategic plan is critical. By investing in your professional growth, the district is also investing in achieving its goals. This is true of every employee at every level within the organization."[10]

Step 3: Identify How Students Will Benefit

The mission of every school focuses on students. If you can tie your professional development opportunity to how it will impact students, you are going to have a higher success probability of getting the funding approved.

Haggerty explained it well, "When student stories are shared to a broader audience to be celebrated, when a levy/bond issue passes providing essential funding to the classrooms, when parents/guardians are made aware of the safety of their child in a crisis or emergency, or when a vital partnership is formed providing outside resources to an initiative supporting students, the benefits are clear. And these are just a handful of examples that can be tied to how communications work helps students directly."[11]

If you can gain the skills you need to be successful in these efforts, there is an obvious impact on students. NSPRA and your state SPRA chapters often have webinars and other resources on this topic. Of course, #SocialSchool4EDU does, too!

Step 4: Identify How Your Job Performance Will Improve

Learning better strategies and gaining access to improved tools will ultimately make you more productive in your work. Be prepared to make a case for how your boss and others in your school will benefit from your increased job performance.

"Just the ability to share ideas is so essential for any profession," Haggerty said. "In this profession, people have always been willing to share their work and their ideas for others to utilize. I think it is one of the most giving professions around. School communications is fast-paced and ever-changing, so not having to re-create from scratch is so helpful."

Social media is always changing, which makes that aspect of your job particularly challenging.

JEN BODE, a development department assistant with Arizona Lutheran Academy, is a member of our social media program. She said, "The biggest advantage of being surrounded by other School PR professionals and social media managers is the motivation the group provides! Need ideas? Go to the group. Got issues? Go to the group. Need moral support? Yep, go to the group. The group has helped me celebrate the positives, navigate the negatives, and wade through the questions of social media."[12]

Step 5: Evaluate a Cost Justification

Often an investment in training can lead to cost savings for your school. For example, supposed you saved time each week doing a specific task because of an investment in training. In that case, you could easily prove a payback for the initial investment.

If a membership program costs $795 for one year, and if the knowledge you gain helps attract or retain just one student in your school, that return back into the district could be as much as $7,000 or more. That means there is nearly a 10X return on the initial investment – WOW!

Jen Bode shared: "Administrators often do not recognize the role that social media and communication plays in retention and future recruitment."[13]

It's true. Your communications efforts have a direct financial impact.

Step 6: Time Your Ask (Maybe)

Your school has a budget cycle. Typically, fiscal years run from July 1 to June 30 each year. If your PD opportunity is thousands of dollars, you're likely going to need to have it in the budget in order for it to be approved. So, get prepared! Budget planning happens throughout the year, so make sure you know the timing for submitting requests.

Timing your ask to the budget cycle is important for bigger ticket items, but smaller costs are not as critical. Every budget has some leeway, and if you've done your homework in steps one through five, the timing of your ask may not be as critical. Are you asking for a few hundred dollars when the payback for students is high? Well then, it's totally worth it for the district!

Step 7: It's Time to Ask in Person!

Don't hide behind your computer, asking via email. You've put together a lot of work to justify your request, so ask your boss for fifteen minutes of their time to talk through this opportunity. You know them better than I do, so you'll know if they want printed materials to go along with your conversation.

Before going into the meeting, strike a power pose. Be confident in your ask, knowing that you did your research and that the funds you are asking for will have a direct impact on students!

I would suggest having a few notes written down from what you gathered in steps one through six. Bring those to the meeting. Your superintendent or supervisor will know that you did your homework before coming in to ask for money. Your script could go something like this:

> I've been looking into programs that can help enhance my skills for the school, and I've identified what I believe is the best value for our investment. The (program name) provides (list benefits) that will help me (list results).

It's time to put this advice to the test. Whether you are looking at the #SocialSchool4EDU membership program or another association or training, it's time to start your homework and make the ask. The worst thing that can happen is a no. And at least they know that you are interested in growing in your role. But remember, if you never ask, the answer is always no.

I'm sure that if you go in confident and prepared, you will earn the funding you deserve.

Is this book helping with your professional development?

If so, share that with others by leaving a review on Amazon or your favorite online bookstore.

YOU'VE GOT THIS!

- [ ] Social media chats, blog posts, and podcasts will keep you informed on changes and developments.
- [ ] Don't underestimate the power of observing and learning from what others are doing.
- [ ] Consider having a mentor; it can be a wonderful way to learn from another professional.
- [ ] Getting PD at the national level is important, but making local connections with like-minded professionals is equally critical.
- [ ] Join NSPRA and your state SPRA.
- [ ] Release cost guilt. An investment in training can lead to cost savings for your school.
- [ ] Put together a plan to make a confident pitch for funding for learning programs.

JOIN "THE CREW" 20

Are you struggling to find the time, content, and support to build engagement on your school's social media pages? In our membership group, we focus on setting you up with systems for success.

Case Study

Meet Dawn Brauner: Dawn is a communication specialist for Portage Community School District in Wisconsin. She has a degree in public relations. She is the one-person communications department for her district.

The challenge: Looking for a place to start, Dawn discovered the free tools available from #SocialSchool4EDU. She started with a minimal social media presence, a disengaged audience, and a load of stress. Dawn soon transformed her district's minimal social media presence into a vibrant, engaged community. She built connected audiences across Facebook, Instagram, and Twitter. Her social media posts began reaching hundreds – and soon, thousands – of people. But Dawn's time is extremely limited. She's pulled in a million different directions every week.

The process: To maximize her limited time, Dawn joined our membership program, Social Media Crew for Schools. Dawn now dedicates minimal time to social media while still maintaining her stellar results. She builds off the ideas shared in the group, attends trainings that fit her schedule and interests, and implements time-saving hacks. If Dawn has a question that she fears is foolish, she comes to a place where judgment is nonexistent and support is abundant: The Crew!

The outcome: Dawn has flourishing, engaged social media channels with a large reach. She receives

CASE STUDY

ongoing tips, tricks, and best practices in a supportive community. The Social Media Crew for Schools is one contributing factor to Dawn's school district's social media presence and success. Dawn said the membership group is her "go-to source" when she hits a roadblock for ideas. "By being a part of the membership group, I feel like I don't have to reinvent the wheel every time I am trying to diversify our social media efforts. I can take ideas . . . add my own touch to them by creating graphics and original photos! Moreover, I can use other members' experiences to answer any questions I may have, and I know that I am not out here alone as a school social media coordinator."[1]

Your mind may be swimming with ideas, and your to-do list is growing after using the resources in this book. Where do you begin? How do you find the support you need? I know that you want to be a rockstar social media manager for your school. To do that, you need training and support – which you've received in this book, but there is still so much more to cover.

Social media should be the last thing that causes you stress. I understand the pressure you feel to run social media effectively, which is why my team at #SocialSchool4EDU has built a community to support you. With our membership program, you will never feel alone or lost about what to share on social media or how to do it.

Social Media Crew for Schools

I invite you to join Social Media Crew for Schools, the #SocialSchool4EDU membership program. We train K–12 school staff to be social media storytellers so you can stand out from other schools, celebrate your students and staff, and reach thousands in your community every day. We are all about getting it done – and getting it done well!

We focus on mastering the basics and creating efficient systems to set you up for success. With millions of school social media posts under our belt, we've seen it all and know how to help. That's why we've created this community to support, train, and inspire school PR professionals on social media.

If you can dedicate thirty to forty-five minutes every week to soaking up the latest social media tips, tricks, and shortcuts, this is the right place for you. If you enjoy self-paced learning, we've got your back. And if you appreciate having a sounding board when making decisions, this community is 100 percent the right place for you.

This program is for anyone involved in social media for your school. You may have the title of communications director or PR specialist, or you may be a principal or executive assistant. We have a lot of media specialists and teachers in the group. I want to be clear that this group is not only for people with the word *communication* in their title.

We've created this community to support, train, and inspire school PR professionals on social media.

Is our program really any different from what you can find for free online? Well, no! We don't share or discuss anything that you couldn't eventually find on your own! The difference is that you'll save countless hours researching since everything you need is in one place. Plus, every person in the group works in school PR, so you have access to the best and brightest minds in our industry.

> "Not having a formal background in communications, I find this group to be my apprenticeship. I learn every day how to tweak what I am doing to promote our school district and in turn helping to educate our staff. Group members are knowledgeable, eager to help, and contribute. I do not feel awkward asking my simplistic questions because others seem to have them too."
>
> —**LORI MARINI**, District Learning Coach, Northern Ozaukee School District, Wisconsin

Is Membership for You?

This program creates results, removes hassle, and implements time-saving hacks. But is it the right fit for you? You can find more details on my website: www.socialschool4edu.com.

If you're looking for advanced training in social media advertising, or if you already have a robust social media strategy in place for your district and aren't looking for new ideas, this program isn't for you.

The Crew, as we love to call it, offers these features:

- ✔ Improves your consistency on your social media platforms
- ✔ Allows you to create content by implementing proven systems
- ✔ Provides the resources and tools you need to improve your results
- ✔ Steps you out of your comfort zone with new ideas to implement
- ✔ Creates a like-minded community so you never feel alone

At #SocialSchool4EDU, we're committed to providing everything you need to create the results you'd like!

HOW "THE CREW" WORKS

Are you ready to become a social media storyteller? We'd love to have you in the community. Here's how it works:

- We provide the support and guidance you need.
- You have access to conversations with the #SocialSchool4EDU team and social media managers from across the country.
- The biggest benefit is community!
- From comments to questions, celebrations, and tips, you'll find reassurance and the right tools so that you are not alone.
- Access to a Canva directory with hundreds of templates specific for schools.
- Use keywords to search for previous posts to find info.
- Monthly Live events – join live or watch later.
- Monthly webinar with an expert.

- Monthly skills session.
- Helpful documents to customize for policies and more.
- A monthly challenge to push you out of your comfort zone.
- Get a weekly email recap of everything that happened in the group with clickable links to every resource from the week.

This is a district membership! So, you and two other people from your school district can join. This membership is meant for public, private, independent, faith-based, and charter schools. And the resources inside of the group can be shared with others inside of your school or district. For example, if you have social media managers at each of your forty-two schools, you could share recorded training sessions or other resources with those people. No extra charge at all!

Membership Program FAQs

Q: What if I don't have the budget for this?
A: Professional development funding is part of every school's budget. If better social media leads to more students attending your school, this program pays for itself.

Q: What if I don't have enough time?
A: It takes less than one hour a week to make substantial progress. And the more tips and tricks you implement, the more time you will save in your day.

Q: My school doesn't think social media is important. What can I tell them?
A: Social media is a powerful tool that can help you reach thousands of people, every day, with positive stories from your school. This has a direct impact on your school's enrollment and ability to pass ballot measures. The bottom line: if you aren't telling your story, who is?

Grab It!

Watch what others are saying about their experience in the Social Media Crew for Schools. And download a one-page flyer to share with your school admins so that you can ask to join. You'll find that link in the bonus resources.

Q: Other training programs haven't worked for me. How is this any different?

A: A one-time social media training will not get you very far. You need ongoing support to keep up with all the changes in social media. You also need a safe place to ask questions and brainstorm ideas with your peers. This is all part of our program.

Q: Can't I find all this information for free?

A: You can certainly find the answers to your social media questions online. But how long does that take you? And is it customized to social media for schools? We do the legwork, curating the best information and helping you apply ideas that will produce immediate results. Inside the membership program, you will also experience facilitated virtual discussions and live trainings that are targeted to your pain points.

Q: What if I already know what I'm doing with social media?

A: If you already know it all, this program isn't for you! But we believe we have something for everyone. We have veteran school communicators with twenty-plus years of experience learning alongside individuals who are brand-new to their roles. Everyone in our program is open to trying new ideas and learning new strategies.

Grab It!

Need more case studies? You'll find links for two in the bonus resources.

Q: You give away so much for free. Why would I pay for more?

A: In the words of one of our members, Jeanne Baudino Berlin, "If all you are seeing is the free stuff . . . you ain't seen nothing yet!"[2]

I've been blogging since 2014, creating YouTube videos since 2015, and cranking out free downloads, webinars, and Q&As along the way. All of this content is specifically aimed at elevating school social media and making it easy (and fun!) for those running it.

But inside the online membership program, you get even more.

Step-by-step guides to help you learn basic and advanced social media concepts. A private community of support with your peers for ideas, advice, and inspiration. 24/7 access to ask questions and troubleshoot problems. Live, facilitated training sessions and access to ALL recorded resources dating back to 2014.

Member Szilvi Lázár had this to say: "Being a member is like having a 5-star hotel smorgasbord breakfast every morning, all-inclusive!!! You can go back to any resource any number of times, search the file directory or the discussions by keyword and you have access to the recordings of ALL previous webinars, skill trainings, everything. If you become a member today it'd be like you'd been a member from the start!"

Member Lori Marini also summed it up well: "The community is the most valuable resource. The consumable resources are wonderful but the personal connections are what keep me coming back."

You'll feel less alone and more supported and empowered than you are, right now. In short, if you've loved our free stuff – we're so glad! If you found value there, imagine how much more you can get with our paid content.

> #SocialSchools4EDU has provided so many great resources to help me systematically build my social media strategy. As a one-person shop, being able to find ideas that are easy to implement and not overwhelming are key!
>
> —**LYNSEY ADMIRE**, Former Director of Communications, Van ISD, Texas

Q: Why would I need to join year after year in this membership program?

A: Let's face it, social media is always changing. In a perfect would, there would be an effective "one and done" training that would prepare you for managing all the ups and downs, but that's just not possible! The team at #SocialSchool4EDU has its finger on the pulse of what's working for schools every day, every month, every quarter, and every year. When you join our membership, you'll always get the most up-to-date information and the latest training! Plus, the 24/7 support of our private community is a safe and valuable place to bounce ideas and get your questions answered.

Pause to Consider

A lack of social media may already be costing you a great deal. We have the tools you need. At #SocialSchool4EDU, we've trained thousands of schools just like yours. Schools that just want to celebrate their students and staff but don't know how.

- ✔ How much is poor use of social media costing you?
- ✔ How many current families have no idea about the success stories happening in your classrooms?
- ✔ How many prospective families aren't even considering your school because they don't see how amazing your staff is or how many opportunities you offer?
- ✔ How much time is your staff wasting on social media strategies that don't work for schools?
- ✔ How many opportunities are you missing to build trust and transparency with your community so they support you on the next ballot question?

We want to provide support so you can stop being overwhelmed and start enjoying your job. With our program, you will never feel alone or lost about what to share on social media or how to do it. You get one year of unlimited, 24/7 access.

YOU'VE GOT THIS!

- ❑ Membership in Social Media Crew for Schools will provide community and support as you get started with practical systems.
- ❑ Stay current with trends, ask questions, and find resources with the affordable membership for up to three people from your school at one price.
- ❑ Find empowerment and encouragement inside the membership group.
- ❑ Low on time? Investing even one hour per week can make a difference.
- ❑ On the fence? Let the testimonials from others show you how they experienced social media success.
- ❑ You can become a social media rockstar!

BONUS Q AND A

Q: How does ADA compliance factor into school social media?

A: Without knowing it, you're possibly excluding students and parents when it comes to your electronic communications and exposing your school district to a possible discrimination claim from the Office of Civil Rights.[1]

Can the visually and hearing-impaired access your PDFs and experience your videos and social media content? Some principles to keep in mind:

- We want to serve our community the best we can because it is the right thing to do, even if the law hasn't yet required compliance in your state.
- We don't want to be motivated only by search engine optimization.
- We want to adhere to accessibility guidelines so that assistive technologies can deliver the content to those who interact with us.

Let's break that down a little more. In a webinar for our membership group, I interviewed several experts on digital inclusion. Some of the tips they shared included:

- Alt text describes what's happening in photos for those who can't see the image. It includes what the people in the photo are doing and their facial expressions. Example: "Crossing guard smiles at a group of students as they follow him to the street." If your post caption includes an accurate depiction of the story in the photo, alt text may or may not be needed.

- In a PDF, add hyperlinks on text that tells people where the link is taking them rather than just a URL.
- A PDF that is only an image cannot be interpreted by a screen reader. In the membership community, we have discussed specific processes for creating accessible PDFs.
- Use accessibility checkers to let you know how accessible your document or image is, including checking color and contrast.
- Include captions on video. Be aware of mistakes in automatic captioning and the importance of editing.

Q: How do I include our virtual school in posts?

A: Whether your district has a full-time virtual school or occasional virtual learning days, photos are a powerful way to showcase the amazing work of students. Ask guardians to submit photos that feature students doing more than staring at a screen. This could include doing an assignment, at home physical education class, saying the Pledge of Allegiance, or practicing an instrument. You can also post about student co-ops and extracurricular events for older students. If you search the *#SocialSchool4EDU* blog for virtual learning, you'll discover some great ideas! As with in-person learning, be sure you have consent for posting.

Q: Where do I find platform-specific tips?

A: Because social media is always changing, I've chosen not to focus on platform-specific how-to info in this book. But we feature it on the *#SocialSchool4EDU* blog often. If you search for a post called "Social Media Best Practices for Schools" on the blog, there's a handy article with the best tips and tricks for the three major social media platforms used by schools: Facebook, Instagram, Twitter.

And even better, we discuss changes and tips often in the membership group. There's a wealth of info waiting for you there with new content every week.

Q: Should I use a Fake Facebook profile?

A: Never use a shared login for a dummy Facebook profile. Always connect real, active personal profiles directly to the page. Facebook is cracking down on "fake" profiles, and you will likely lose access to your page if you violate this principle. Facebook continues to make it easier to assign users to your page in a secure way. Make sure every admin on your page has two-factor authentication turned on!

Q: Should I boost posts or run paid ads?

A: Social media advertising involves paying money to have your content reach more social media users. You can do this on all platforms. I personally don't have any expertise in this area, but schools around the country have had success with enrollment campaigns, teacher recruitment posts, and more. You'll want to weigh costs and reach, comparing print or billboard advertising with targeted ads on social media, to decide which is most effective for you. There are plenty of training programs out there that will teach you about social media advertising and also agencies that specialize in school communication.

••

Do you have more questions? We answer a lot more in the membership community. Join The Crew to have a place to ask questions any time!

celebrate YOUR SCHOOL

ONE STORY AT A TIME!

NOTES

Chapter 1

1. Andrea Gribble, "What Happens When School Leaders Embrace Social Media?, *#SocialSchool4EDU* (blog), https://socialschool4edu.com/what-happens-when-school-leaders-embrace-social-media/.
2. Andrea Gribble and guest Jerry Almendarez, "Creating a Successful Brand Ambassador Program with Jerry Almendarez," episode 135, *Mastering Social Media for Schools* (podcast), November 7, 2022, https://socialschool4edu.com/podcast/135/.
3. Content in this section first appeared on Andrea Gribble, "What to Say When People are Against Using Social Media for Your School District," *#SocialSchoopl4EDU* (blog), September 17, 2019, https://socialschool4edu.com/what-to-say-when-people-are-against-using-social-media-for-your-school-district/.
4. "Social Media Fact Sheet," Pew Research Center, April 7, 2021, accessed October 17, 2022, https://www.pewresearch.org/internet/fact-sheet/social-media/.
5. Hannah Feller, "Feature My School: Piqua City Schools," *#SocialSchoopl4EDU* (blog), June 17, 2019, https://socialschool4edu.com/feature-my-school-piqua-city-schools/.
6. Hannah Feller, "Feature My School."
7. Hannah Feller, "Feature My School."
8. Inspired by a quote from George Couros on Twitter, "We need to make the positive so loud that the negative becomes almost impossible to hear," https://twitter.com/gcouros/status/660840853754134528?lang=en.
9. "Mobile Fact Sheet," Pew Research Center, April 7, 2021, https://www.pewresearch.org/internet/fact-sheet/mobile/.
10. "Social Media Fact Sheet," Pew Research Center, April 7, 2021, https://www.pewresearch.org/internet/fact-sheet/social-media/.
11. John Gramlich, "10 Facts about Americans and Facebook," Pew Research Center, June 1, 2021, https://www.pewresearch.org/fact-tank/2021/06/01/facts-about-americans-and-facebook/.
12. John Gramlich, "10 Facts."
13. Andrea Gribble and guest Dr. Joe Sanfelippo, "Changing the Narrative on Education with Dr. Joe Sanfelippo," *Mastering Social Media for Schools* (podcast), episode 7, May 25, 2020, https://socialschool4edu.com/podcast/7.
14. Andrea Gribble and guest Dr. Joe Sanfelippo, "Changing the Narrative."

Chapter 2

1. Andrea Gribble with guest Lana Snodgras, "How to Manage All the Things in a One-Person SchoolPR Shop with Lana Snodgras," *Mastering Social Media for Schools* (podcast), episode 71, August 16, 2021, https://socialschool4edu.com/podcast/71.
2. New Auburn School District Facebook page, Details About New Auburn School District, https://www.facebook.com/newauburnschool/about_details.
3. West Plains School District, Social Media Rules of Engagement and Posting Guidelines, accessed November 30, 2022, https://www.zizzers.org/Page/6238.
4. Excerpts from this section first appeared at: Andrea Gribble, "Staff Social Media Guidelines for Personal Use," *#SocialSchool4EDU* (blog), August 18, 2020, https://socialschool4edu.com/staff-social-media-guidelines-for-personal-use/.
5. Andrea Gribble, "Staff Social Media Guidelines for Personal Use," *#SocialSchool4EDU* (blog), August 18, 2020, https://socialschool4edu.com/staff-social-media-guidelines-for-personal-use/.
6. "Social Media Guidelines for Personal Use," Dothan City Schools, https://www.dothan.k12.al.us/site/handlers/filedownload.ashx?moduleinstanceid=3790&dataid=5982&FileName=Dothan%20City%20Schools%20Social%20Media%20Guidelines%20for%20Personal%20Use.pdf/.
7. Andrea Gribble, "Staff Social Media Guidelines."
8. Excerpts from this section first appeared at: Andrea Gribble, "Creating Your School Social Media Policy," *#SocialSchool4EDU* (blog), January 13, 2016, https://socialschool4edu.com/creating-your-school-social-media-policy/.
9. Family Educational Rights and Privacy Act (FERPA), U.S. Department of Education, https://www2.ed.gov/policy/gen/guid/fpco/ferpa/index.html/.
10. Excerpts from this section first appeared at: Andrea Gribble, "Should I Share Student Names on Posts?," *#SocialSchool4EDU* (blog), May 6, 2019, https://socialschool4edu.com/should-i-share-student-names-on-posts/.

Chapter 3

1. From a recorded interview with Andrea Gribble and Brendan Schneider, December 2022.
2. Andrea Gribble, "Social Media Goal Setting for 2020-2021," *Mastering Social Media for Schools* (podcast), episode 12, June 29, 2020, socialschool4edu.com/podcast/12.
3. Barb Nichol, APR, "Are Your Social Media Posts Meeting Your Communication Goals?," *#SocialSchool4EDU* (blog), April 20, 2020, https://socialschool4edu.com/are-your-social-media-posts-meeting-your-communication-goals/.
4. Andrea Gribble, "How to Reach More People with your Facebook Posts (Without Paying for It)," YouTube (video), May 14, 2019, https://youtu.be/1O9gsMn-0TA.
5. Barb Nichol, APR, "Are Your Social Media Posts Meeting Your Communication Goals?" *#SocialSchool4EDU* (blog), April 20, 2020, https://socialschool4edu.com/are-your-social-media-posts-meeting-your-communication-goals/.

6. Dorreen Dembski, "Referendum Communication Guide," *#SocialSchool4EDU* (blog), October 15, 2017, https://socialschool4edu.com/referendum-communication-guide/.
7. Dorreen Dembski, "Referendum Communication."
8. Andrea Gribble, "Your Referendum To-Do List for Social Media," *#SocialSchool4EDU* (blog), September 11, 2018, https://socialschool4edu.com/your-referendum-to-do-list/.

Chapter 4

1. From a recorded interview in 2022.
2. Excerpts from this section first appeared at: Andrea Gribble, "Why You Need ONE District-wide Facebook Page," *#SocialSchool4EDU* (blog), October 2, 2017, https://socialschool4edu.com/one-district-wide-facebook-page/.
3. Andrea Gribble, "Dos and Don'ts for School Staff Running Their Own Social Media Pages," #SocialSchool4EDU (blog), February 9, 2020, https://socialschool4edu.com/dos-and-donts-for-school-staff-running-their-own-social-media-pages/.
4. Excerpts from this section first appeared at: Andrea Gribble, "8 Social Media New Year's Resolutions for 2022," *#SocialSchool4EDU* (blog), January 18, 2022, https://socialschool4edu.com/8-social-media-new-years-resolutions-for-2022/.
5. Andrea Gribble and guest Angela Brown, "What the Latest Data Reveals About Parent Behavior and School Marketing Priorities with Angela Brown," *Mastering Social Media for Schools* (podcast), episode 137, November 21, 2022, https://socialschool4edu.com/podcast/137.
6. Andrea Gribble, "Social Media Is Like a Free Puppy, Not a Free Beer," *#SocialSchool4EDU* (blog), July 22, 2022, https://socialschool4edu.com/social-media-is-like-a-free-puppy-not-a-free-beer/.
7. From a Word document shared by Lana Snodgras.

Chapter 5

1. Andrea Gribble and guest Megan Anthony, "Rebranding, Social Media, & Anxiety – Oh My! With Megan Anthony," *Mastering Social Media for Schools* (podcast), episode 55, April 26, 2021, https://socialschool4edu.com/podcast/55.
2. Andrea Gribble, "Branding Tips for Schools: Part 1," *#SocialSchool4EDU* (blog), September 20, 2022, https://socialschool4edu.com/branding-tips-for-schools-part-1/.
3. Andrea Gribble, "Branding Tips for Schools: Part 1."
4. Andrea Gribble, "Branding Tips for Schools: Part 1."
5. Andrea Gribble, "Branding Tips for Schools: Part 1."
6. Andrea Gribble and guest Ian Halperin, "A Conversation with a School PR Veteran – Ian Halperin," *Mastering Social Media for Schools* (podcast), episode 76, September 20, 2021, https://socialschool4edu.com/podcast/76.

7. Heidi Feller, "Branding Tips for Schools: Part 2," *#SocialSchool4EDU* (blog), September 30, 2022, https://socialschool4edu.com/branding-tips-for-schools-part-2/.
8. Heidi Feller, "Branding Tips for Schools: Part 2.
9. Heidi Feller, "Branding Tips for Schools: Part 2.
10. Heidi Feller, "Branding Tips for Schools: Part 2.
11. Heidi Feller, "Branding Tips for Schools: Part 2.

Chapter 6

1. Andrea Gribble and guest Marissa Weidenfeller, "It's All About That Brand with Marissa Weidenfeller," *Mastering Social Media for Schools* (podcast), episode 61, June 7, 2021, https://socialschool4edu.com/podcast/61.
2. This list first appeared at: Andrea Gribble, "Using Creative Writing to Drive Engagement on Social Media," *#SocialSchool4EDU* (blog), November 27, 2019, https://socialschool4edu.com/using-creative-writing-to-drive-engagement-on-social-media/.
3. Andrea Gribble and guest Patricia Weinzapfel, "Building Strong Parent Partnerships with the Power of Our Words with Patricia Weinzapfel, MS," *Mastering Social Media for Schools* (podcast), episode 113, June 6, 2022, https://socialschool4edu.com/podcast/113/.
4. Adapted from a live skills session led by school communicator, writer, strategist, and content creator Kristin Boyd Edwards, in the #SocialSchool4EDU Social Media Membership Program.
5. Andrea Gribble and guest Erin McCann, "Meet Your New BFF in #SchoolPR with Erin McCann," *Mastering Social Media for Schools* (podcast), episode 67, July 19, 2021, https://socialschool4edu.com/podcast/67.

Chapter 7

1. Andrea Gribble and guest Sherese Nix, "#The GISD Effect & the Importance of Storytelling with Sherese Nix," *Mastering Social Media for Schools* (podcast), episode 111, May 23, 2022, https://socialschool4edu.com/podcast/111/.
2. Portions of content in this chapter first appeared at: Andrea Gribble, "#Hashtag Questions – Answers for Schools," *#SocialSchool4EDU* (blog), February 25, 2016, https://socialschool4edu.com/hashtag-questions-answers-for-schools/.
3. A version of this content first appeared at: Andrea Gribble, "#Hashtag Questions – Answers for Schools," *#SocialSchool4EDU* (blog), February 25, 2016, https://socialschool4edu.com/hashtag-questions-answers-for-schools/.
4. Craig Olson, "Telling Our Story #LikeACane," *#SocialSchool4EDU* (blog), August 28, 2017, https://socialschool4edu.com/positive-community-impact/.
5. A version of this list first appeared at: Andrea Gribble, "Recipe for an Awesome School #Hashtag," *#SocialSchool4EDU* (blog), June 30, 2015, https://socialschool4edu.com/recipe-for-an-awesome-school-hashtag/.

6. This list first appeared at: Andrea Gribble, "15 Hip #Hashtags for Schools," *#SocialSchool4EDU* (blog), October 9, 2019, https://socialschool4edu.com/15-hip-hashtags-for-schools/.

Chapter 8

1. "Be Seen. Be Heard. Be Bold. Branding for SoWashCo Schools," CEL website, accessed December 3, 2022, https://www.celpr.com/sowashco-schools-branding/.
2. "The SoWashCo Schools Brand," South Washington County Schools website, accessed December 3, 2022, https://www.sowashco.org/brand.
3. This content originally appeared at: Andrea Gribble, "It's Time to Update Your Cover Photos!," *#SocialSchool4EDU* (blog), September 14, 2016, https://socialschool4edu.com/time-update-cover-photos/.
4. Andrea Gribble, "Social Media – The Profile Picture Debate," *#SocialSchool4EDU* (blog), August 8, 2915, https://socialschool4edu.com/social-media-the-profile-picture-debate/.
5. This section excerpted from: Allison Martinson, "Social Media Graphics: Just OK is Not OK," *#SocialSchool4EDU* (blog), April 22, 2019, https://socialschool4edu.com/social-media-graphics-just-ok-is-not-ok/.
6. Andrea Gribble and guest Maci Stover, "Facebook Ads, Branding, and Classroom Visits with Maci Stover," *Mastering Social Media for Schools* (podcast), episode 120, July 25, 2022, https://socialschool4edu.com/podcast/120/.
7. From a recorded interview in 2022.
8. Andrea Gribble and guest Elishia Seals, "Putting Your Heart into Social Media (& Some Canva Ninja Tips) with Elishia Seals," *Mastering Social Media for Schools* (podcast), episode 109, May 9, 2022, https://socialschool4edu.com/podcast/109/.

Chapter 9

1. Shared during Social Media Bootcamp – Week 1 – Accountability Session, December 2, 2022 in the private membership group for #SocialSchool4EDU.
2. Parts of this section first appeared at: Andrea Gribble, "Back to School – Checklist for Getting Staff Involved in Social Media," *#SocialSchool4EDU* (blog), July 7, 2020, https://socialschool4edu.com/back-to-school-checklist-for-getting-staff-involved-in-social-media/.
3. Andrea Gribble, "Back to School – Checklist for Getting Staff Involved in Social Media."
4. Andrea Gribble with guest Sarah O'Donnell, "Falling Back in Love With Your Job in School PR with Sarah O'Donnell," *Mastering Social Media for Schools* (podcast), episode 131, August 16, 2021, https://socialschool4edu.com/podcast/131/.
5. Andrea Gribble with guest Sarah O'Donnell, "Falling Back in Love With Your Job in School PR with Sarah O'Donnell."

6. Parts of this section first appeared at: Andrea Gribble, "How to Create a Social Media Directory for Your School," *#SocialSchool4EDU* (blog), November 29, 2021, https://socialschool4edu.com/how-to-create-a-social-media-directory-for-your-school/.
7. Andrea Gribble, "How to Create a Social Media Directory."
8. Andrea Gribble, "How Lunch 'n Learns Can Boost Staff Social Media Participation," *#SocialSchool4EDU* (blog), August 5, 2022, https://socialschool4edu.com/how-lunch-n-learns-can-boost-staff-social-media-participation/.

Chapter 10

1. Andrea Gribble and guest Callen Moore, "Called to do 'Moore' in a One-Person Communications Office with Callen Moore," *Mastering Social Media for Schools* (podcast), episode 128, September 19, 2022, https://socialschool4edu.com/podcast/128/.
2. These strategies first appeared at: Andrea Gribble, "Social Media Made Easy – 9 Strategies for Generating Awesome Content," *#SocialSchool4EDU* (blog), July 29, 2019, https://socialschool4edu.com/social-media-made-easy-9-strategies-for-generating-awesome-content/.
3. Andrew Hutchinson, "Instagram Stories is Now Being Used by 500 Million People Daily," Social Media Today (website), January 31, 2019, https://www.socialmediatoday.com/news/instagram-stories-is-now-being-used-by-500-million-people-daily/547270/.
4. Andrea Gribble and guest Nicole Valles, "BOLD Social Media Advice that Works with Nicole Valles," *Mastering Social Media for Schools* (podcast), episode 89, December 20, 2021, https://socialschool4edu.com/podcast/89.
5. Content in this section first appeared in a guest post: Heidi Feller, "The Tech Divide: An Easy Tip to Get More Photos for Social Media," *#SocialSchool4EDU* (blog), August 6, 2021, https://socialschool4edu.com/the-tech-divide-an-easy-tip-to-get-more-photos-for-social-media/.
6. Heidi Feller, "The Tech Divide."
7. Portions of this section and the 52 questions first appeared at: Andrea Gribble, "52 Questions That Will Increase Engagement on Facebook!," *#SocialSchool4EDU* (blog), July 16, 2021, https://socialschool4edu.com/52-questions-that-will-increase-engagement-on-facebook/.

Chapter 11

1. Andrea Gribble and guest Nicole Valles, "BOLD Social Media Advice that Works with Nicole Valles," *Mastering Social Media for Schools* (podcast), episode 89, December 20, 2021, https://socialschool4edu.com/podcast/89.
2. Parts of this chapter first appeared at: Andrea Gribble, "How to Find Student Stories Your Community Will Love!," *#SocialSchool4EDU* (blog), December 10, 2021, https://socialschool4edu.com/how-to-find-student-stories-your-community-will-love/.

3. Lakeside School District Facebook post, November 17, 2021, accessed August 12, 2022. https://www.facebook.com/LakesideSD/posts/4652949401485044/.
4. Parts of this chapter first appeared at: Andrea Gribble, "How to Find Student Stories Your Community Will Love!," *#SocialSchool4EDU* (blog), December 10, 2021, https://socialschool4edu.com/how-to-find-student-stories-your-community-will-love/.
5. Facebook post from Oostburg School District, October 1, 2021, accessed August 12, 2022, https://www.facebook.com/oostburgschools/posts/4544082492280298/.
6. Andrea Gribble, "Don't Go Looking for a Fight! Why Some Information Should Not Be Posted on Social Media," #SocialSchool4EDU (blog), October 26, 2021, https://socialschool4edu.com/dont-go-looking-for-a-fight-why-some-information-should-not-be-posted-on-social-media/.
7. Shared in a recorded interview with Andrea Gribble.
8. Parts of this section first appeared at: Andrea Gribble, "Calendar of Celebrations for Schools," *#SocialSchool4EDU* (blog), July 1, 2019, https://socialschool4edu.com/calendar-of-celebrations-for-schools/.
9. Parts of this section first appeared at: Andrea Gribble, "Calendar of Celebrations for Schools."

Chapter 12

1. Andrea Gribble and guest Amanda Keller, "Tales of an Accidental School Social Media Professional with Amanda Keller," *Mastering Social Media for Schools* (podcast), episode 104, April 4, 2022, https://socialschool4edu.com/podcast/104/.
2. Excerpts from this section first appeared at: Andrea Gribble, "How to Use "Batching" to Increase Productivity in Your School PR Role," *#SocialSchool4EDU* (blog), January 27, 2022, https://socialschool4edu.com/how-to-use-batching-to-increase-productivity-in-your-school-pr-role/.
3. Andrea Gribble, "How to Use "Batching" to Increase Productivity in Your School PR Role," *#SocialSchool4EDU* (blog), January 27, 2022, https://socialschool4edu.com/how-to-use-batching-to-increase-productivity-in-your-school-pr-role/.
4. Stephanie Sinz, "5 Social Media Features for Your School," *#SocialSchool4EDU* (blog), November 12, 2019, https://socialschool4edu.com/5-social-media-features-for-your-school/.
5. Stephanie Sinz, "5 Social Media Features."
6. Stephanie Sinz, "5 Social Media Features."
7. Content in this section first appeared at: Andrea Gribble, "Student Spotlights – Creating this Weekly Feature is EASY!," *#SocialSchool4EDU* (blog), January 15, 2018, https://socialschool4edu.com/student-spotlights/.
8. Stephanie Sinz, "5 Social Media Features."
9. Stephanie Sinz, "5 Social Media Features."
10. Stephanie Sinz, "5 Social Media Features."

11. Stephanie Sinz, "5 Social Media Features."
12. Emily Rae Schutte, "What I Learned This Week" – An Easy Social Media Feature!," *#SocialSchool4EDU* (blog), October 27, 2020, https://socialschool4edu.com/what-i-learned-this-week-an-easy-social-media-feature/.
13. Emily Rae Schutte, "'What I Learned This Week' – An Easy Social Media Feature!"
14. Content in this section first appeared at: Andrea Gribble, "#TBT Guide – What to Look For in Yearbooks," *#SocialSchool4EDU* (blog), April 20, 2016, https://socialschool4edu.com/tbt-guide-look-yearbooks/.
15. Stephanie Sinz, "5 Social Media Features."
16. Stephanie Sinz, "5 Social Media Features."

Chapter 13

1. Andrea Gribble and guest Jason Wheeler, "TikTok for Schools & How to Turn Students into Storytellers with Jason Wheeler," *Mastering Social Media for Schools* (podcast), episode 29, October 26, 2020, https://socialschool4edu.com/podcast/29/.
2. As quoted in Heidi Feller, "Student Contributors for Social Media," *#SocialSchool4EDU* (blog), March 12, 2019, https://socialschool4edu.com/student-contributors-for-social-media/.
3. Heidi Feller, "Student Contributors."
4. Heidi Feller, "Student Contributors."
5. As quoted in Heidi Feller, "Student Contributors for Social Media."
6. LeAnne Bugay, "Need help with social media? Hire a student intern!," *#SocialSchool4EDU* (blog), January 19, 2021, https://socialschool4edu.com/need-help-with-social-media-hire-a-student-intern/.
7. As quoted in Heidi Feller, "Student Contributors."
8. Andrea Gribble and guest Christine Paik, "Engaging Students as Content Creators with Christine Paik," *Mastering Social Media for Schools* (podcast), episode 74, September 6, 2021, https://socialschool4edu.com/podcast/74/.
9. Richmond Public Schools, "Instagram Takeover Toolkit: Fine Arts Friday," March 2022.

Chapter 14

1. Adapted from a live skills session led by school communicator, writer, strategist, and content creator Kristin Boyd Edwards, in the #SocialSchool4EDU Social Media Membership Program.
2. Hannah Feller, "Top Ten Ways to Start a Post," *#SocialSchool4EDU* (blog), March 26, 2018, https://socialschool4edu.com/top-ten-ways-start-post/.
3. Hannah Feller, "Top Ten Ways to Start a Post."
4. Adapted from a live skills session led by school communicator, writer, strategist, and content creator Kristin Boyd Edwards, in the #SocialSchool4EDU Social Media Membership Program.

5. A version of this list first appeared at: Andrea Gribble, "Using Creative Writing to Drive Engagement on Social Media," *#SocialSchool4EDU* (blog), November 27, 2019, https://socialschool4edu.com/using-creative-writing-to-drive-engagement-on-social-media/.
6. Hannah Feller, "How to Polish a Post," *#SocialSchool4EDU* (blog), April 23, 2018, https://socialschool4edu.com/how-to-polish-a-post/.
7. Inspired by guest post from Hannah Feller, "Basic Grammar and Style for Social Media," *#SocialSchool4EDU* (blog), January 1, 2018, https://socialschool4edu.com/basic-grammar-style-social-media/.
8. Andrea Gribble and guest Erin McCann, "Meet Your New BFF in #SchoolPR with Erin McCann," *Mastering Social Media for Schools* (podcast), episode 67, July 19, 2021, https://socialschool4edu.com/podcast/67.
9. Adapted from a live skills session led by school communicator, writer, strategist, and content creator Kristin Boyd Edwards, in the#SocialSchool4EDU Social Media Membership Program.
10. Stephanie Sinz, "How Adding Emojis Can Engage Your Social Media Audience!," *#SocialSchool4EDU* (blog), June 3, 2019, https://socialschool4edu.com/how-adding-emojis-can-engage-your-social-media-audience/.
11. Stephanie Sinz, "How Adding Emojis Can Engage."
12. Stephanie Sinz, "How Adding Emojis Can Engage."
13. "20 Calls to Action to Drive engagement on Social Media," #SocialSchool4EDU, PDF document.

Chapter 15

1. From a recorded interview in 2022.
2. Excerpts in this section first appeared at: Andrea Gribble, "5 Tips for Responding to Negative School Social Media Comments," *#SocialSchool4EDU* (blog), January 21, 2016, https://socialschool4edu.com/5-tips-to-for-responding-to-negative-school-social-media-comments/.
3. A flow chart example shared on www.socialschool4edu.com was inspired by the process shared in: Kristin Magette, *Embracing Social Media: A Practical Guide to Manage Risk and Leverage Opportunity* (Lanham, Maryland: Rowman & Littlefield Publishers, 2014).
4. Excerpts from this section first appeared at: Andrea Gribble, "How to Handle the Community Gossip Groups," *#SocialSchool4EDU* (blog), September 9, 2021, https://socialschool4edu.com/how-to-handle-the-community-gossip-groups-on-social-media/.
5. Andrea Gribble, "Community Gossip Groups."
6. Excerpts in this section first appeared at: Andrea Gribble, "Schools Under Attack," *#SocialSchool4EDU* (blog), March 12, 2018, https://socialschool4edu.com/schools-under-attack/.
7. Excerpts from: Andrea Gribble, "Community Gossip Groups."
8. Andrea Gribble, "Community Gossip Groups."

9. Excerpts from: Andrea Gribble, "Community Gossip Groups."
10. Andrea Gribble, "Community Gossip Groups."
11. Andrea Gribble, "Community Gossip Groups." (Comment updated for this book.)
12. Excerpts from: Andrea Gribble, "Community Gossip Groups."
13. Andrea Gribble, "Community Gossip Groups."
14. Andrea Gribble, "Schools Under Attack," *#SocialSchool4EDU* (blog), March 12, 2018, https://socialschool4edu.com/schools-under-attack/.

Chapter 16

1. From a recorded interview with Andrea Gribble and Laraine Weschler in December 2022.
2. Excerpts in this section first appeared at: Andrea Gribble, "Video: The Key to More Engagement," *#SocialSchool4EDU* (blog), July 21, 2016, https://socialschool4edu.com/video-key-engagement/.
3. This list first appeared at: Andrea Gribble, "Creating Better Videos with Tools You Already Have," *#SocialSchool4EDU* (blog), October 26, 2016, https://socialschool4edu.com/creating-better-videos/.
4. Andrea Gribble and guest Jessica James, "How to Use Facebook Life to Showcase Your Students with Jessica James," *Mastering Social Media for Schools* (podcast), episode 94, January 24, 2022, https://socialschool4edu.com/podcast/94/.
5. Andrea Gribble and guest Jessica James, "How to Use Facebook Life to Showcase Your Students with Jessica James."
6. Andrea Gribble and guest Morgan Delack, "Create a Facebook Live Show in Your District with Morgan Delack," *Mastering Social Media for Schools* (podcast), July 12, 2021, https://socialschool4edu.com/podcast/66/.
7. Andrea Gribble and guest Morgan Delack, "Create a Facebook Live Show."

Chapter 17

1. From a recorded interview with Andrea Gribble and Jenny Starck, December 2022.
2. Excerpts in this section first appeared at: Andrea Gribble, "How to Create a Social Media Report Card in Less Than One Hour!," *#SocialSchool4EDU* (blog), January 8, 2021, https://socialschool4edu.com/how-to-create-a-social-media-report-card-in-less-than-one-hour/.
3. Andrea Gribble and guest Amanda Keller, "Tales of an Accidental School Social Media Professional with Amanda Keller," *Mastering Social Media for Schools* (podcast), episode 104, April 4, 2022, https://socialschool4edu.com/podcast/104/.
4. Andrea Gribble, "How to Create a Social Media Report Card in Less Than One Hour!," *#SocialSchool4EDU* (blog), January 8, 2021, https://socialschool4edu.com/how-to-create-a-social-media-report-card-in-less-than-one-hour/.

Chapter 18

1. K12prWell, www.k12prwell.org.
2. From a recorded interview December 9, 2022.
3. Content in this chapter first appeared at: Andrea Gribble, "10 Practical Tips to Avoid Burnout in #SchoolPR," *#SocialSchool4EDU* (blog), October 27, 2020, https://socialschool4edu.com/10-practical-tips-to-avoid-burnout-in-schoolpr/.
4. Andrea Gribble, "10 Practical Tips."
5. Stephen R. Covey, The 7 Habits of Highly Effective People (New York: Simon & Shuster, 2020).
6. This list first appeared at: Andrea Gribble, "10 Practical Tips to Avoid Burnout in #SchoolPR."
7. Andrea Gribble and guest Kristin Magette, "Wellness Check-in for School Leaders with Kristin Magette, APR," *Mastering Social Media for Schools* (podcast), episode 20, August 24, 2020, https://socialschool4edu.com/podcast/20/.
8. Andrea Gribble, "No – You Can't Do it All! A Story of School PR Balance," *#SocialSchool4EDU* (blog), September 19, 2019, https://socialschool4edu.com/no-you-cant-do-it-all/.

Chapter 19

1. From a recorded interview in 2022.
2. Andrea Gribble and guest Janet Swiecichowski, "How to Use Facebook Life to Showcase Your Students with Jessica James," *Mastering Social Media for Schools* (podcast), August 30, 2021, https://socialschool4edu.com/podcast/73/.
3. National School Public Relations Association, accessed November 4, 2022, https://www.nspra.org/About-Us.
4. Andrea Gribble and guest Cara Adney, "Professional Development: Good for your Skills & Good for your Soul with Cara Adney," *Mastering Social Media for Schools* (podcast), episode 27, October 12, 2020, https://socialschool4edu.com/podcast/27/.
5. This content originally appeared at: Andrea Gribble, "Professional Development Opportunities for School Communicators," *#SocialSchool4EDU* (blog), September 30, 2021, https://socialschool4edu.com/professional-development-opportunities-for-school-communicators/.
6. Andrea Gribble, "Professional Development Opportunities for School Communicators," *#SocialSchool4EDU* (blog), September 30, 2021, https://socialschool4edu.com/professional-development-opportunities-for-school-communicators/.
7. Andrea Gribble, "Professional Development Opportunities."
8. Andrea Gribble and guest Susan Brott, "Mentorship – It's All About Learning with Susan Brott, APR," *Mastering Social Media for Schools* (podcast), episode 77, September 27, 2021, https://socialschool4edu.com/podcast/77/.

9. This content first appeared at: Andrea Gribble, "7 Steps to Landing Professional Development Funding for Your School PR Role," *#SocialSchool4EDU* (blog), June 23, 2021, https://socialschool4edu.com/7-steps-to-landing-professional-development-funding-for-your-school-pr-role/.
10. Shane Haggerty in Andrea Gribble, "7 Steps to Landing Professional Development Funding."
11. Shane Haggerty in Andrea Gribble, "7 Steps to Landing Professional Development Funding."
12. Jen Bode in Andrea Gribble, "7 Steps to Landing Professional Development Funding."
13. Jen Bode in Andrea Gribble, "7 Steps to Landing Professional Development Funding."

Chapter 20

1. Hannah Feller, "Center Stage: Making an Impact with a One-Person Communications Department," guest post on *#SocialSchool4EDU* (blog), August 8, 2019, https://socialschool4edu.com/center-stage-portage-community-school-district/.
2. "Membership Program," #SocialSchool4EDU (website), https://socialschool4edu.com/services/membership-program/.

Bonus Q&A

1. Andrea Gribble (guests Bryce Neiman and Nicole Lyons), "Digital Inclusion – Making Communications Accessible to All," webinar from #SocialSchool4EDU, September 15, 2021, private video link.

ACKNOWLEDGMENTS

Whoa. I can't believe I get the chance to acknowledge all of the people who helped make this book happen! I'm afraid I'm going to leave someone out, but I'll give it my best shot.

Bill – You are my rock. You were the one who encouraged me to pursue this business when I wanted to give up. You have cheered me on, lifted me up, and loved me through it all. I am so blessed to do life with you!

My girls – Aliya and Kyra. It's been a wild ride, but I wouldn't change a thing. From the difficult decision to sell our house, to living with Grandma and Grandpa, to the green house in Sand Creek, to the MLB player's log cabin, to our beautiful lake home – you were troopers through it all. Starting a business when you're a single mom is not easy, but you are the reasons I worked so hard. I am so proud of the young women you are growing up to be, and I appreciate that you have always believed in me.

My boys – Chase, Kolton, Grant, and Garon. You four make your dad and me so proud. You provide constant entertainment and make vacations much more memorable. My next book might just be real life short stories of the Gribble boys!

Brian Henning – Thank you for believing in me before I believed in myself. You were the one who reached out for help and introduced me to the need for social media help for schools. I am so thankful for the opportunity.

The New Auburn School District – The students and staff in little New Auburn trusted me with their stories when I launched with them on April 15, 2014. I didn't really know what I was doing, but everyone got excited about sharing the stories from our little one-building school district in the middle of Wisconsin. You have a lot to be proud of, and I'm so thankful to be the one who gets to share those stories with the world on Facebook, Instagram, and Twitter.

Joe Sanfelippo – Thanks for making the time for a thirty-five-year-old single mom to come in and learn about the incredible storytelling happening in Fall Creek. Your willingness to share, your excitement about the work being done in your school, and your encouragement that schools needed a service like mine helped me commit to making this business work.

Kristin Magette – To the first "celebrity" I met in NSPRA . . . Thank you for replying to my tweet when I was reading your book! You showed me the power of social media and how it can lead to genuine connections. I consider you a mentor and friend, and I can't believe that I've written a book now, too!

Sarah Stokes Herzog – To the one who helped me rediscover the JOY in my life and in my business! Without our powerful year together, I don't know if this book would have happened so fast. Your caring heart, listening ear, and encouraging soul have lifted me to a higher purpose, and I am forever grateful!

Michelle Rayburn – I never would have been able to create such an amazing book without your help! When I was overwhelmed, you broke down the steps into manageable tasks. You took my words that I've been sharing for nearly a decade and created a valuable guide to help my friends in school communication.

Emily Schutte – Another secret to my success is you, Miss Emily! I am so thankful for the passion, knowledge, and expertise you bring into this business. Your words and marketing genius have helped put a little Wisconsin girl on the map when it comes to school social media.

Mom and Dad – You're the reason I'm here! Thank you for always encouraging me. You have supported me through the years and really had to step up when I started this business. Thanks for sharing your home with me and the girls. I also appreciate how you're my biggest fans. Dad, I know that you love me and are proud of me!!

Sharon Gribble and Bill Gribble Sr. – You brought the best gift in the world to me – my husband, Bill. It's been a crazy ride, but your love and support have meant the world to us!

My #SocialSchool4EDU team – I don't do anything alone. The team of amazing women that surround me daily are truly my superpower. Steph, Heidi, Allison, Andrea, Angie, and Melissa – along with all of the account managers – I am so thankful that you trusted to join me on our mission to celebrate students and connect communities.

Membership Crew – I truly appreciate your commitment to celebrating your schools! You provide continuous insight and ideas that help guide my work, and I couldn't do what I do without you.

All of my #SchoolPR and NSPRA friends – Thanks for welcoming me into your world. I was an outsider. I didn't even know school PR existed. And now you've taught me nearly everything I know. You are the most genuine, caring people on the planet. The work you do matters every day. It is a blessing to serve you and call you friends!

ABOUT #SOCIALSCHOOL4EDU

Andrea's passion is helping schools recognize their daily awesomeness and sharing that story with the world. She's built a team that celebrates hundreds of schools across the country! #SocialSchool4EDU provides full social media management and runs a vibrant online community that provides ongoing professional development for school social media champions. Andrea also hosts the *Mastering Social Media for Schools* weekly podcast.

socialschool4edu.com

I'm a natural-born cheerleader. I love cheering people on, helping them do their best, and shining a spotlight on their accomplishments. In high school, I cheered from the sidelines, but in my career, I celebrate through the use of technology. Helping schools recognize their daily awesomeness and sharing that story with the world is my thing! But it didn't start out that way.

After a successful thirteen-year run in the corporate world, I found myself on the sidelines – but this time, I had been benched. I was laid off in 2013. As a single mom with two little girls, I had to become my own cheerleader.

It wasn't easy, but I taught myself the ins and outs of social media marketing. The superintendent at my small school district in Wisconsin asked if I could help launch Facebook and other social channels. He didn't have time to figure it out and – luckily! – he trusted me enough to help. (Hey, I did graduate at the top of my high school class back in 1996. Who cares if we only had thirty-two kids in my grade?)

I loved it. It didn't feel like marketing. My job was to share photos, videos, and stories on Facebook, Twitter, Instagram, and YouTube. I got to be a cheerleader again – this time, celebrating the amazing students and staff at that district!

And guess what? That little community in New Auburn started reaching more than 5,000 people every week with the stories we shared. Word started to get around, and I picked up a second school, and then a third.

Today, I've built an entire cheerleading squad that helps me celebrate hundreds of schools across the country! We provide full social media management, and we have a vibrant online community that provides ongoing professional development for school social media champions. I also host a weekly podcast called *Mastering Social Media for Schools* and have a variety of free resources available on my website, socialschool4edu.com.

When I'm not cheering on K–12 schools (public, private, charter, online – we serve them all!), I'm cheering on my family. I have an amazing husband, two daughters, and four stepsons. Between basketball, volleyball, golf, and football, we also squeeze in as much time on the lake as possible.

Andrea Gribble

Facebook @SocialSchool4EDU
Instagram @socialschool4edu
LinkedIn @andreagribble
Twitter @andreagribble
YouTube @SocialSchool4EDU

Made in the USA
Coppell, TX
13 October 2024

38574068R00174